10.99

Mastering

Statistics

Macmillan Master Series

Accounting
Advanced English Language
Advanced Pure Mathematics
Arabic
Banking
Basic Management
Biology
British Politics
Business Administration
Business Communication
Business Law
C Programming
Catering Theory
Chemistry
COBOL Programming
Communication
Databases
Economic and Social History
Economics
Electrical Engineering
Electronic and Electrical Calculations
Electronics
English as a Foreign Language
English Grammar
English Language
English Literature
English Spelling
French
French 2

German
German 2
Human Biology
Italian
Italian 2
Japanese
Manufacturing
Marketing
Mathematics
Mathematics for Electrical and
  Electronic Engineering
Modern British History
Modern European History
Modern World History
Pascal Programming
Philosophy
Photography
Physics
Psychology
Science
Social Welfare
Sociology
Spanish
Spanish 2
Spreadsheets
Statistics
Study Skills
Word Processing

Macmillan Master Series
Series Standing Order ISBN 0–333–69343–4
You can receive future titles in this series as they are published by placing a standing order. Please contact your bookseller or, in case of difficulty, write to us at the address below with your name and address, the title of the series and the ISBN quoted above.
Customer Services Department, Macmillan Distribution Ltd
Houndmills, Basingstoke, Hampshire RG21 6XS, England

Mastering

Statistics

Third Edition

Tim Hannagan

MACMILLAN

First edition 1982
Reprinted three times
Second edition 1986
Reprinted five times
Third edition 1997
Published by
MACMILLAN PRESS LTD
Houndmills, Basingstoke, Hampshire RG21 6XS
and London
Companies and representatives
throughout the world

ISBN 0–333–66057–9

A catalogue record for this book is available
from the British Library.

This book is printed on paper suitable for recycling and
made from fully managed and sustained forest sources.

10   9   8   7   6   5   4   3   2
06   05   04   03   02   01   00   99   98

Typeset by Wearset, Boldon, Tyne and Wear
Printed in Hong Kong

To Yvonne, Jane, Kathleen, Liam, Kevin

Contents

Preface   xi
Acknowledgement   xiii

1   Statistical information   1
  1.1   Definition of statistics   1
  1.2   Statistical information   1
  1.3   Numerical data   2
  1.4   The abuse of statistics   2
  1.5   The use of statistics   4
  1.6   Primary and secondary data   5
  1.7   Secondary statistics   7
  1.8   Aids to calculation   8

2   The accuracy of statistical information   12
  2.1   Approximation   12
  2.2   Levels of tolerance   13
  2.3   Error   14
  2.4   Rounding   16
  2.5   Absolute error   18
  2.6   Relative error   18
  2.7   Biased, cumulative or systematic error   19
  2.8   Unbiased or compensating error   19
  2.9   Calculations involving approximation and error   20
  2.10   Spurious accuracy   24

3   Collecting statistical information   26
  3.1   Surveys   26
  3.2   Survey methods   27
  3.3   Observation   29
  3.4   Interviews   31
  3.5   Questionnaires   33
  3.6   Why sample?   33

3.7	Advantages of sampling	35
3.8	Objectives of sampling	37
3.9	The basis of sampling	37
3.10	Sampling errors	38
3.11	Sample size	39
3.12	Sample design	39
3.13	Bias in sampling	40
3.14	Sampling methods	41
4	**How to present statistics (1)**	**52**
4.1	The aims of presentation	52
4.2	Tables	53
4.3	Classification	53
4.4	Frequency distributions	57
4.5	Reports	59
4.6	Histograms	59
4.7	Frequency polygons	63
4.8	Frequency curves	65
5	**How to present statistics (2)**	**68**
5.1	Bar charts	68
5.2	Pie charts	71
5.3	Comparative pie charts	73
5.4	Pictograms	74
5.5	Comparative pictograms	75
5.6	Cartograms or map charts	76
5.7	Strata charts	76
5.8	Graphs	77
5.9	Semi-logarithmic graphs	81
5.10	Straight-line graphs	82
5.11	Gantt charts	83
5.12	Break-even charts	84
5.13	The Z chart	85
5.14	The Lorenz curve	86
5.15	Presentation and perception	89
6	**Summarising data: averages**	**92**
6.1	The role of the average	92
6.2	The arithmetic mean	94
6.3	The median	100
6.4	The quartiles	103
6.5	The mode	106
6.6	The geometric mean	109
6.7	The harmonic mean	110
7	**Summarising data: dispersion**	**112**
7.1	Dispersion	112

7.2	The range	114
7.3	The interquartile range	116
7.4	The standard deviation	118
7.5	The variance	127
7.6	The coefficient of variation	127
7.7	Conclusions	128

8 Statistical decisions 130
8.1	Estimation	130
8.2	Probability	131
8.3	The normal curve	136
8.4	Probability and the normal curve	139
8.5	Standard error	140
8.6	Tests of significance	142
8.7	Confidence limits	144
8.8	Statistical quality control	145
8.9	Conclusion	146

9 Comparing statistics: index numbers and vital statistics 148
9.1	Index numbers	148
9.2	Weighted index numbers	150
9.3	Chain-based index numbers	153
9.4	Problems in index-number construction	153
9.5	The Index of Retail Prices	155
9.6	Vital statistics	155

10 Comparing statistics: correlation 160
10.1	What is correlation?	160
10.2	Scatter diagrams	162
10.3	Correlation tables	163
10.4	The product moment coefficient of correlation	165
10.5	Coefficient of rank correlation	167
10.6	Spurious correlation	168
10.7	Regression	170

11 Trends and forecasting 171
11.1	Trends and forecasting	171
11.2	Time series	172
11.3	Moving averages	173
11.4	Seasonal variations	174
11.5	Irregular and residual fluctuations	179
11.6	Linear trends	180
11.7	Conclusions	183

12 Mastering more statistics 186
| 12.1 | More statistics | 186 |
| 12.2 | Statistical proof | 187 |

12.3 Experiments   189
12.4 More inductive statistics   192
12.5 More methods of collecting information   196
12.6 Checklists and ratings   198
12.7 Scaling   199
12.8 Financial statistics   205
12.9 Information technology   209
12.10 The next step   211

Appendixes   213
A1 Further reading   215
A2 Area table   216
A3 Basic mathematics   217
A4 How to pass statistics examinations   231

Index   236

Preface

The aims of this book are to provide a complete statistical foundation for a wide variety of readers, some of whom may have very limited mathematical knowledge. The book includes a section on simple arithmetic as a reminder of basic mathematical techniques.

The book is designed to provide a comprehensive appreciation of statistical sources, concepts, methods and applications and therefore it provides a strong base for further quantitative and numerical study.

It aims at stressing the relevance of statistics to everyday life and in a wide range of business situations. The book is designed for:

- the *student* who is studying for a range of examinations, courses and modules;
- the *general reader* who wants to have a greater understanding of events, facts and figures in everyday life;
- the *businessperson* who needs to understand, interpret and make use of statistical information and to use statistical techniques in industry and commerce.

This book provides a broad coverage of the statistics required in the General Certificate of Secondary Education (GCSE) syllabus and for General National Vocational Qualification (GNVQ) and professional examinations. This third edition has been revised to ensure that the chapters relevant to recent developments are up to date. Learning objectives have been added for each chapter and the presentation of material has been developed to help understanding and revision. The last chapter provides a summary of statistical tests, to add to the discussion on inductive statistics and increases the depth of study on methods of collecting information. It suggests a range of applications for statistics, areas for further study and includes a section on the implications of information technology. The list of formulae, a glossary of the main statistical terms and the reading list have been updated. The section on 'How to pass statistics examinations' provides help and advice to examination candidates.

The book is an introduction to statistics for students on social science, public administration and many other degree and diploma courses. It provides for aspects of basic numeracy for programmes such as the GNVQ, NVQ, and management and professional courses.

Assignments are included at the end of each chapter to reinforce the information in the chapter. These assignments are designed to provide for both individual learning and group activity. While it is quite possible to read and understand the book without completing any of these assignments, they are aimed at illustrating the practical uses of a particular technique or concept as well as providing practice for examinations. Further practice can be obtained from the questions and exercises (and answers) in *Work Out Statistics* (Tim Hannagan, Macmillan, second edition, 1996) which is complementary to this book in its approach and coverage.

On completion of this book the reader should be aware of the main statistical sources of information, understand numerical concepts and be proficient in basic statistical methods. The reader should also be able to apply a variety of statistical techniques to help solve business and everyday problems.

TIM HANNAGAN

Acknowledgement

The author and publishers wish to thank the Controller of Her Majesty's Stationery Office for permission to use tables from *Social Trends*.

1 Statistical information

OBJECTIVES

- To define and describe 'statistics'
- To analyse the nature of statistical information and numerical data
- To consider the use and abuse of statistics

1.1 Definition of statistics

Statistics is concerned with scientific methods for collecting, organising, summarising, presenting and analysing data, as well as drawing valid conclusions and making reasonable decisions on the basis of this analysis.

Data (the singular 'datum' is not often used) are things known or assumed as a basis for inference. Statistics is concerned with the systematic collection of numerical data and its interpretation.

The word 'statistics' is used to refer to:

- numerical facts, such as the number of people living in a particular town;
- the study of ways of collecting and interpreting these facts.

It can be argued that figures are not facts in themselves; it is only when they are interpreted that they become relevant to discussions and decisions.

1.2 Statistical information

Everybody collects, interprets and uses information, much of it in a numerical or statistical form. People receive large quantities of information every day through conversations, television, computers, the radio, newspapers, posters, notices and instructions. It is just because there is so much information available that people need to be able to absorb, select and reject it, to sift through facts to pick out what is interesting and useful. In everyday life as well as in business and industry, certain statistical information is necessary and it is important to know where to find it or how to collect it.

As consumers, everybody has to compare prices and quality before making a decision about what goods to buy. As employees, people want to compare wages and conditions of work. As employers, firms want to control costs and expand profits.

The main function of statistics is to provide information which will help in making decisions. It can be argued that to make decisions without the help of statistics is asking for trouble; rather like asking a doctor to diagnose a patient's illness without the doctor being able to collect information on the patient's condition, such as temperature, pulse, blood pressure and so on. Statistics provides this type of information by providing a description of the present, a profile of the past and an estimate of the future.

1.3 Numerical data

The ease of processing numerical data has increased its use and availability; however, there is often a tendency to avoid numerical information in favour of written, visual or verbal information. (It is interesting to notice that it is considered unacceptable to be illiterate, but not to be innumerate.) One of the interesting exceptions to the avoidance of numerical data is in sport, where tables and figures are accepted as normal. People have little difficulty in understanding batting averages or pools coupons. Perhaps this indicates that everyone is capable of understanding complicated numerical information if they are sufficiently well motivated.

Many facts can be given sensibly only in numerical form. This includes fairly complicated data, for example financial information or information in a football league table (Table 1.1). If addition, subtraction, multiplication and division present few difficulties, and you can understand Table 1.1, then mastering statistics will present few problems. (Basic mathematical rules are outlined in Appendix A3.)

1.4 The abuse of statistics

Statistics are used to support a case, either to make it appear 'scientific', or because collecting and examining all the statistical evidence relevant to a case will increase its validity. However, statistics can sometimes be used in misleading ways.

You might prove anything by figures. (CARLYLE)
There are three kinds of lies: lies, damn lies, and statistics. (DISRAELI)
Don't be a novelist – be a statistician, much more scope for the imagination. (MEL CALMEN)
He uses statistics as a drunken man uses a lamp post – for support rather than illumination. (ANDREW LANG)

PRESENTING NUMERICAL DATA

Tables can be a clear and concise way of presenting statistical information. It would require several closely typed pages to describe all the information given in Table 1.1, and to analyse the information would fill several more pages. Note, for example, that, although the link between winning games and position in the league table is close, it is not exact. In order to obtain their 46 points, Arsenal won 50% of their games while Sheffield Wednesday, with the same number of points, won only 44% of their games. Sheffield Wednesday, however, drew 37% of their games while Arsenal drew only 14% of theirs and this difference in the number of draws evens up the points earned. Sheffield Wednesday is above Arsenal in the table because of a more favourable goal difference.

Table 1.1   A football league table.

Football club	Games				Goals		Points
	Played	Won	Drawn	Lost	For	Against	
Everton	27	17	5	5	60	31	56
Tottenham	27	16	6	5	53	27	54
Man. Utd	28	14	7	7	51	32	49
Liverpool	28	13	9	6	40	23	48
Sheff. Wed	27	12	10	5	42	26	46
Arsenal	28	14	4	10	48	37	46
Wimbledon	28	13	7	8	38	33	46
Leeds	27	14	3	10	41	35	45
Chelsea	28	10	10	8	42	33	40
Aston Villa	27	10	7	10	38	41	37
Blackburn	27	10	7	10	33	37	37
Newcastle	29	9	9	11	43	53	36

These quoted statements are both right and wrong; the use of statistics is a matter of interpretation. **Interpretation is a matter of judgement based on knowledge:** for example, the knowledge of what is meant by the term 'inflation'. It is at least as important that a figure or the result of a calculation is understood as it is that it is accurate. There is little point in arriving at a correct answer to a calculation if it is not known what it means. However, accurate statistical information can be used selectively. For example, if the objective is to show that the rate of inflation is low, it would be possible to say that it is running at $\frac{1}{2}$% on the basis that prices rose by this amount last month. If the objective is to show that the level of inflation is high, it would be possible to state that inflation is running at 18%, on the basis that the average level of price rises over the last ten years has been 18%.

In fact, it would be accurate to say that 'you can prove nothing with statistics'. Figures are used to support theories, opinions and prejudices. It can be argued that in the above examples it was decided first of all that the objective was to show that the rate of inflation was low or high and then statistics were found to support these decisions.

Ideally, the collection and analysis of statistics are carried out as accurately and objectively as possible. In practice, statistics are at times employed to sensationalise, inflate, confuse or oversimplify. Therefore it is important to be realistic about the way in which statistics are collected and summarised.

1.5 The use of statistics

The principal function of statistics is to narrow the area of disagreement which would otherwise exist in a discussion and in that way help in decision making. Statistics can be a stabilising force, dispelling rumour and uncertainty, helping to solve arguments arising from individual cases or circumstances by providing a factual foundation to debates and decisions which would otherwise be dominated by subjectively based theories and opinions.

The grandmother theory

This theory runs on the lines of 'my Grandmother (or Grandfather) smoked 50 cigarettes (drank a bottle of whisky/never did any exercise/ate like a horse) every day and lived to be 105 years old; therefore smoking (drinking, lying around, over-eating) cannot be bad for you'. In a sense statistics are the opposite of the 'grandmother theory'. Statistics do not prove that everybody who smokes heavily dies young, but the statistical evidence does show that people are more likely to die young if they smoke heavily than if they do not.

It is this feature of systematic collection of data that distinguishes statistics from other kinds of information. **Statistics provide a method of systematically summarising aspects of the complexities of social and economic problems.**

It is useful to divide statistics into:

- **descriptive statistics,** including the presentation of data in tables and diagrams, as well as calculating percentages, averages, measures of dispersion and correlation, in order to display the salient features of the data and to reduce it to manageable proportions;
- **inductive statistics,** involving methods of inferring properties of a population on the basis of known sample results. These methods are based directly on probability theory.

In fact most knowledge is 'probability knowledge' in the sense that it is only possible to be absolutely certain that a statement is true if the statement is of a restricted kind. Statements such as 'I was born after my father', 'a black cat is black', 'one plus one equals two' are tautologies or disguised definitions, and although we are certain they are true they are of limited value in providing information. Most information is based on sample data rather than on a complete survey of a population, and probability theory is the basis of the areas of statistics that are not purely descriptive (in particular see Chapters 3 and 8).

> The emphasis in the work of the statistician has shifted from a backward-looking process to an analysis of current affairs, proposed future operations and their consequences. Every business organisation and government department uses statistics in its daily work. There has been an enormous development in the collection, processing and storing of statistical information in recent years, which has been assisted by the growth of computer technology. H. G. Wells suggested that 'statistical thinking will one day be as necessary for efficient citizenship as the ability to read and write'. Perhaps that day has come.

Many people have computers and pocket calculators and if they do not need them for business or for study they may still use them in their role as consumers or tax-payers or sports enthusiasts. The use of electronic means of processing data has removed much of the repetitive drudgery of statistical calculations, but has made it more than ever necessary to understand statistical material. Facts and figures cannot be accepted on their 'face value'; it is necessary to know something about the methods used to collect the data and to summarise it in order to understand its imperfections and strengths. Basic mathematics are covered in Appendix A3.

1.6   Primary and secondary data

Numerical data can be divided into primary and secondary data.

Primary data are collected by or on behalf of the person or people who are going to make use of the data. This collection of data involves all the survey and sample methods described in Chapter 3. Once data have been collected, processed and published they become secondary data.

Secondary data are used by a person or people other than the people by whom or for whom the data were collected. They are 'second-hand' data. Few people collect their own statistical material except on a small scale and almost everybody collects some statistical material on a small scale, although not necessarily very systematically. Consumers compare prices at a number of shops before making a purchase; firms collect data about aspects of their business, such as output and sales. However, for many general purposes and for collections of large-scale statistical material and for material collected systematically, most people use tabulations produced by others.

Because secondary data are second-hand, it is important to know as much as possible about them, just as it is important to know as much as possible about a second-hand car before buying it. It is useful to know how the data have been collected and processed in order to appreciate the reliability and full meaning of the statistics. Tables can be misleading: for example, a table may show that a firm has increased its output of a small component this year, compared with last year, by 100%. This result appears less impressive if it is known that the firm produced only 5 of these components last year and 10 this year. If these numbers had been 5000 and 10 000, the result would have been more impressive and the percentage more appropriate.

The main points to be considered in the use of secondary data are:

- how the data has been collected,
- how the data has been processed,
- the accuracy of the data,
- how far the data has been summarised,
- how comparable the data is with other tabulations,
- how to interpret the data, especially when figures collected for one purpose are used for another.

Generally, with secondary data, people have to compromise between what they want and what they are able to find. However, there are great advantages in using secondary data:

1   it is cheap to obtain – many government publications are relatively cheap and libraries stock quantities of data produced by the government, by companies and other organisations;
2   large quantities of secondary data are easily available – although they may not be exactly what is wanted;
3   there is a great variety of data on a wide range of subjects;
4   much of the secondary data available has been collected for many years and therefore it can be used to plot trends.

Secondary data is of value to:

- *the government* – to help in making decisions and planning future policy;
- *business and industry* – in areas such as marketing and sales, in order to appreciate the general economic and social conditions, and to provide information on competitors;
- *research organisations* – by providing social, economic and industrial information.

Published statistics tend either to be the product of an 'administrative process' or to come from specially conducted enquiries or surveys. Statistics on unemployment are a good example of the former. They are collected mainly as part of the process of registering people as unemployed so that they are eligible for unemployment benefit. Therefore the statistics are collected as part of an administrative process and are not entirely appropriate for other uses.

Unemployment statistics are used to estimate the number of people unemployed, but they are not an accurate record of the actual number of people not in work because large numbers of people who are looking for work do not register as unemployed. This includes people who are not eligible for benefit, such as housewives, and people who expect to find work at any moment and do not bother to register. Also, the unemployment statistics provide no information on underemployment: that is, people who are at work but are not working at full capacity. Therefore, at best, official unemployment statistics provide a considerable underestimation of the level of unemployment.

Statistics from opinion polls, the Census of Population and the Family Expenditure Survey are good examples of surveys and enquiries designed solely to obtain information. The Census of Population in the UK is carried out every ten years in an attempt to discover a wide range of information on households. This information may help to formulate administrative decisions but it is not collected as part of an administrative process.

Surveys and enquiries can more easily be used as secondary data than the bi-product statistics produced as part of an administrative process. Not only has the data been collected for the purpose of providing information but also details are available describing how the survey or census has been conducted.

1.7 Secondary statistics

Statistics compiled from secondary data are termed *secondary statistics*. Therefore calculations made on the basis of government figures on, for example, unemployment are described as secondary statistics. The tables of unemployment figures on which the calculations are based are secondary data.

The problems of using secondary data apply equally to secondary statistics. Anyone using published statistics needs to consider the purpose for which they were originally compiled. Secondary statistics are common because secondary data is so easily available. There is a mass of secondary sources of data in official statistics, private publications and through computer networks. In the process of monitoring their business activities, firms accumulate statistics which are mainly for their own benefit but are also used as a base for official sources through the various returns firms make to government departments. Firms can compare their own statistics with those available for their industry and for the economy.

Businesses use statistics to provide clear and concise summaries of the

firm's business activity, to carry out surveys and samples to produce trends and forecasts, which can provide the basis for decisions.

Governments produce statistics because:

- the data is part of the administrative process;
- they want to be able to measure the effects of their policies;
- they want to monitor the effects of external factors on their policies;
- they need to be able to assess trends so that they can plan future policies.

1.8   Aids to calculation

Electronic calculators

These first came into common use in the early 1960s. Calculators are of many different kinds, but a relatively inexpensive pocket calculator will perform all the calculations needed for basic statistics. All calculators will add, subtract, multiply and divide. Most calculators now include a wide range of facilities such as, for example, a square root key, while the more sophisticated ones enable advanced mathematical and scientific calculations to be made. For business purposes it may be useful for a calculator to have these attributes:

- For frequent calculations of such things as Value Added Tax (VAT), a constant factor facility is useful. This allows the multiplication or division of a series of numbers by the same number without having to enter each calculation separately.
- A facility to calculate percentages, mark-ups and discounts directly.
- A memory facility may be useful for complicated calculations.
- A clear-last-entry facility allows a mistake to be corrected in an entry without having to go all the way back to the beginning of the calculation.

In Japan the traditional abacus is still used for teaching numeracy. This is a compact counting frame (or soroban) which is considered to help in providing mastery of basic mathematical concepts, while it can be argued that electronic calculators only provide answers to calculations and not an understanding of the basic mathematical processes involved in obtaining the answers.

Personal computers

Computers have all the advantages of electronic calculators with very much more power: they are faster, they have a memory and they can be programmed to carry out a range of activities. They can, for example, provide not only the answer to a calculation but also a step-by-step learning programme to analyse the way in which the answer has been achieved, which may help in the development of a conceptual framework for numeracy (see also Section 12.9).

Logarithms

For 250 years before electronic calculators came into everyday use, people used logarithms to help them carry out laborious calculations. The word comes from the Greek for 'calculating with numbers'.

The use of logarithms avoided long multiplication and division calculations, substituting the relatively simpler addition and subtraction processes. Logarithms were inexpensive and faster than manual calculation, but they have been superseded by calculators and computers.

Logarithms are exponents and are based on the same principles as exponents. Exponents provide a shorthand method of writing out multiple multiplication; they are the symbol showing the power of a factor. For example:

$$2^6 = 2 \times 2 \times 2 \times 2 \times 2 \times 2 = 64$$

6 is the exponent providing the instruction to multiply 2 by itself 6 times.

In common logarithms the logarithm of a number is the exponent of 10 that will produce that number.

LOGARITHMS

$$10^0 = 1 \qquad 10^{-1} = \tfrac{1}{10} = 0.1$$
$$10^1 = 10 \qquad 10^{-2} = \tfrac{1}{100} = 0.01$$
$$10^2 = 100 \qquad 10^{-3} = \tfrac{1}{1000} = 0.001$$

Therefore the log of:

1 = 0	0.1	= −1
10 = 1	0.01	= −2
100 = 2	0.001	= −3
1000 = 3	0.0001	= −4

Therefore the logarithm of all numbers between

1	and	10	will be 0 plus a fraction
10	and	100	will be 1 plus a fraction
100	and	1000	will be 2 plus a fraction
1	and	0.1	will be −1 plus a fraction
0.1	and	0.01	will be −2 plus a fraction
0.01	and	0.001	will be −3 plus a fraction

Section 5.9 describes the use of semi-logarithmic graphs.

ASSIGNMENTS

1 Discuss the information contained in Table 1.2. What are the problems of using the information from a table of this kind?

2 Discuss (in writing or verbally in a group) what is meant by the term 'information'. Find and describe an example that illustrates the problem of the interpretation of data.

3 What are the main problems involved with using secondary data? Illustrate these problems by reference to any table of secondary data.

4 Consider the kind of problems a person who cannot read or write will face. How far are these problems similar to those of a person who cannot count or understand figures?

5 How far is it possible to agree with the statements quoted at the beginning of Section 1.4? Find examples to support these statements.

6 Outline and discuss the type of statistical information gathered by any one organisation in the normal process of its work. Consider how this information could be used to help the organisation in the future.

7 Research television viewing figures and write a report on the following questions.
   (i)   How many hours of television does the average child and the average adult view each week?
   (ii) Do children view more television than adults, and men more than women?
   (iii) What influence does unemployment and retirement have on average weekly viewing hours?

Table 1.2 Direct and indirect taxes as percentages of gross domestic product (GDP): international comparison.

	Direct and indirect taxes as percentages of GDP			Direct taxes as percentages of GDP		
	Year 1	*Year 5*	*Year 10*	*Year 1*	*Year 5*	*Year 10*
Denmark	27.6	39.8	39.4	15.1	24.6	24.4
Finland	25.3	29.0	34.1	11.8	14.8	20.8
Norway	27.4	32.1	34.1	13.6	14.6	16.9
Sweden	29.6	32.1	33.6	18.0	19.7	21.9
Netherlands	26.0	28.9	31.9	16.2	18.3	21.3
Luxembourg	19.4	20.5	28.6	12.0	13.4	18.9
Ireland	19.7	23.8	28.2	5.9	8.8	11.4
Belgium	21.9	24.0	28.0	9.9	12.3	16.9
Austria	24.4	25.5	26.5	11.5	12.2	13.3
Germany	22.3	23.7	25.6	12.6	13.9	16.7
United Kingdom	21.5	23.5	25.5	12.3	14.1	16.8
New Zealand	17.5	19.9	24.1	10.8	13.0	17.2
Switzerland	15.6	16.8	21.6	9.8	11.1	16.0
France	18.9	19.5	21.0	5.9	6.5	8.5
Australia	15.7	17.2	20.9	8.5	10.2	13.4
Italy			18.7			9.1
USA	14.9	17.3	17.2	10.2	12.2	12.5
Greece	15.3	16.6	16.2	4.8	5.9	5.4
Portugal	9.8	12.4	15.5	2.5	3.5	5.4
Japan	9.7	10.1	10.5	5.2	6.1	7.1
Spain	9.8	9.3	10.0	3.1	3.6	5.0

Source: adapted from *Social Trends* 10

2 The accuracy of statistical information

OBJECTIVES

- To provide an understanding of the accuracy of statistical information
- To define and describe statistical error
- To provide an appreciation of spurious accuracy

2.1 Approximations

Perfect accuracy in statistical information is possible only in limited circumstances. 'One plus one equals two' is perfectly accurate, but in applied areas of activity where problems have to be solved and decisions made, perfect accuracy of this kind is impossible, and therefore approximations are used.

Statistics is built upon approximations of one kind or another. There is very rarely an exact result to be derived from a statistical investigation. To the extent that statistics can be described as a social product rather than an objective method it is concerned with the level of accuracy required at a particular time.

The accountant keeps track of every penny because both sides of the ledger (debit and credit) must balance to the last penny. The statistician may round figures to the nearest unit, or hundred or thousand, depending on the level of accuracy required in a piece of information in order for it to be useful in solving problems and making decisions.

Even in areas where accuracy appears to be a minor problem it may be difficult to obtain. All scientific measurements are to some degree inaccurate, either because no measuring device can record an exact reading or because of the experimenter's inability to read the index accurately. Even exact mathematical results may be only approximate in relation to reality, since they are based on assumptions of a continuity of the circumstances assumed to exist for the purpose of the calculation. If the circumstances alter, so may the true value alter.

APPROXIMATIONS

Example 1

Export figures in the UK are based, not on actual sailings of ships with cargo, but on exporters' returns received by the Department of Trade during the year. Therefore the figures depend on the accuracy of the returns. It was discovered that, at various times, British export figures had been underestimated by millions of pounds a month because small exporting companies had not made the necessary returns.

Example 2

The instructions on some bottles of aspirin state: 'Take 1 or 2 tablets 3 or 4 times a day'. One tablet, 3 times a day would be 3 tablets in total; 2 tablets, 4 times a day would be a total of 8 tablets. The level of tolerance here is considerable. The instructions could be rewritten to say 'Take between 3 and 8 tablets at intervals during the day depending on how you feel'. That is as accurate as they need to be. On the other hand some medicines have to be taken with so much accuracy that they can be applied only under the controlled conditions of a hospital ward.

Example 3

A machine operator may be asked how long he spent maintaining his machine during recent weeks. His answer might be 'five minutes' one week, 'half-an-hour' another week and 'just over an hour' the third week. In fact the times might have been 'three minutes', 'thirty-two minutes' and 'sixty-eight minutes' respectively, but the approximations make more sense in answering a general enquiry which does not require a high degree of accuracy. The answer might have been different depending on the reason for the enquiry and who was making it.

NATIONAL CENSUS

If a national census takes place next week, by the time the census figures are available, the demographic population statistics will have altered because of the births and deaths which have occurred in the intervening period. In fact, although at a specific moment the population of a country is a finite figure, the actual figure available will always be an approximation because of the impossibility of counting every single person alive at that moment.

2.2   Levels of tolerance

The degree of accuracy required in statistics depends upon the type of data being measured and the uses to which it will be put. For every measurement there will be a level of tolerance beyond which inaccuracy becomes unacceptable, and within which inaccuracy is acceptable.

Statistical tables usually contain approximations because tables are a summary of the collected data and are unlikely to contain all the information that was in the original survey. In summarising the data, rounding is almost certain to take place. In most cases this is indicated by putting the degree of approximation in brackets: '(thousands)' would indicate that the column of figures is rounded to the nearest thousand. If the first figure in the column is 31 000, then the true figure would lie between 30 500 and 31 500.

Close approximations are often good enough for the purposes for which the data is required; statistics help to define the limits within which such approximations function. It is important that approximations should be clearly identified. In the same way inaccuracy must be allowed for in statistical investigation by recognising that it exists and that it can be kept within tolerable limits.

2.3 Error

'Error' means the difference between what is acceptable as a true figure and what is taken for an estimate or approximation.

Error in this sense does not mean a mistake. In the calculation $5 + 2 - 1 = 4$ the answer is wrong and therefore in mathematical terms it could be described as an error or a mistake. In statistical terms an 'error' is the difference between the approximate figure and the true figure.

ERROR

Example 1
The size of a crowd at an open-air meeting may be estimated to be 20 000. Nobody knows exactly how many people attended the meeting either because nobody counted exactly or/and because nobody had been able to count the numbers because of constant movement with people arriving and leaving. However, at a particular moment there would have been an exact number in the crowd. If everybody had been made to stand still for long enough it would have been possible to count the numbers there. This would have been the 'true

figure'. If this had been say 18 000, this figure could be compared with the approximate or estimated number of 20 000. The difference, 2000, is the error. This could be put in a different way by saying that the error was 10%.

Example 2
When the population of the UK is said to be 56 million, this does not mean exactly 56 million, but is more likely to mean 56 million to the nearest million. This means that the population could be anywhere between 55.5 million and 56.49 million. The figure of 56 million could be as much as half a million people different from the 'true' figure. Therefore the 56 million could be written as 56 million $\pm \frac{1}{2}$ million. The $\frac{1}{2}$ million is the degree of error involved in the approximation.

The error in Example 1 above is not a 'mistake'. The size of crowds at open-air meetings are often estimated because of the cost, time and difficulty of arriving at an exact figure. Also, exact figures are not usually required.

However, it is important that the crowd size quoted in this example should be qualified in one way or another to indicate that it is an estimate and not an exact figure. The estimated crowd size could be described as 'about 20 000', 'estimated to be 20 000' or '20 000 approximately'.

A mistake is usually involuntary and something to be avoided, while approximations are made to improve the presentation of figures and in this sense statistical error is deliberate. Error is a way of qualifying the degree of accuracy of a result (see also Section 8.5).

Most statistics reflect the degree of accuracy required rather than the degree of accuracy that is possible.

ACCURACY

If it had been thought necessary to measure the size of the crowd at the open-air meeting to an accuracy of say 2%, then a barrier could have been erected, turnstiles installed and everyone counted as they entered the meeting. This would be expensive and time consuming and there would need to be a good reason to make it worthwhile.

Even if there is an incentive to collect accurate data there may still be some error involved. Crowds entering football grounds have to pass through turnstiles because they have to pay, and the numbers have to be counted because the stadium will have a limited capacity. So figures are issued for crowd sizes at football matches which could be completely accurate. In fact, however, even under these conditions there is likely to be some error involved. The counting might be inaccurate at the turnstiles through human or mechanical failure, and some people may have climbed over the walls.

The widespread use of approximations means that any conclusions drawn from figures are themselves subject to error; therefore it is useful to know and to be able to identify the main types of approximations used.

2.4 Rounding

To the nearest whole number

Fractions and decimals are frequently rounded 'to the nearest whole number'. The convention is to round 0.5 and above to the next highest whole number and up to 0.499 (recurring) to the next lowest whole number. Therefore:

6.5 would become 7
6.499 would become 6

In the same way 65 would become 70 if rounded to the 'nearest whole ten' and 64.99 would become 60; 650 would become 700 rounded to the 'nearest hundred' and 649.99 would become 600.

Survey figures are frequently rounded, because if some of the data is not very accurate there is little point in recording other figures with great accuracy, unless they are in some way independent.

NEAREST WHOLE NUMBER

In a survey of petrol sold at service stations, recorded figures may be rounded to the nearest full litre or tens of litres, on the basis that there is likely to be some inaccuracy due to such factors as spillage and evaporation. This means making a decision as to how to round the actual figures. The choices are as follows.

1   To round up to the next highest whole unit, say to the next ten litres. An actual figure of 123 litres would become 130 litres recorded on the survey form.
2   To round down to the next lowest ten litres. An actual figure of 123 litres would be recorded as 120 litres.
3   To round to the nearest ten litres. An actual figure of 123 litres would become 120 litres, and an actual figure of 125 litres would become 130 litres.

Rounding up (1) and rounding down (2) are not recommended in statistics unless there are very good reasons, because they give rise to biased error (see Section 2.7). If it was felt that there was always a wastage of petrol in the above example, so that the amount sold was always less than the amount recorded as going through the petrol pumps, there might be a justification for rounding down. However, rounding to the nearest unit (3) usually means that the small amounts added tend to balance out the small amounts subtracted, so that the final total is close to the true total.

Very large figures become more comprehensible if rounded. To be told that the population of the UK is about 56 million is easier to comprehend than to be given the figure 55 873 451, even if there was some way of knowing that the second figure was accurate. Even if this figure was accurate it would usually be unnecessary to quote it in full in most statistical presentations.

To the nearest even number

Another procedure which can be used in an attempt to reduce bias from rounding is to round a number so that the digit preceding the final zeroes in the approximate value is even and not odd.

> **NEAREST EVEN NUMBER**
>
> If 125 is to be rounded to two significant figures (see below), and to the nearest even number, it would be rounded down to 120, while 135 would be rounded up to 140.

In most circumstances this procedure does not have any advantages over rounding to the nearest whole number.

By truncation

Truncation is a similar procedure to rounding. It consists of the omission of the unwanted final digits.

> **TRUNCATION**
>
> 15.268 truncated to four figures becomes 15.26; truncated to two figures it would become 15.

In some currencies, banks ignore very small denominations of coinage, and pocket calculators truncate any digits lying outside their display capacity.

This procedure produces a downwards bias in the results obtained.

By significant figures

Rounding to a number of significant figures is a process by which the number of digits that are significant are stated and, after that number, zeroes replace other digits.

SIGNIFICANT FIGURES

Calculated figure	Four significant figures	Three significant figures	Two significant figures
213.73	213.7	214	210
0.003 726	0.003 726	0.003 73	0.0037
2 482 731	2 483 000	2 480 000	2 500 000
30 000	30 000	30 000	30 000
20 518	20 520	20 500	21 000

Notice that the zeroes which only indicate the place value of the significant figures (tens, hundreds, thousands) are not counted as significant digits.

Frequently, significant figures are the digits that carry real information and are free of spurious accuracy (see Section 2.10).

2.5 Absolute error

Absolute error is the actual difference between an estimate or approximation and the true figure.

ABSOLUTE ERROR

One person may expect to spend £10 on her shopping, but actually spends £12.50. The absolute error is £2.50. Another person expects to spend £20 on his shopping, but actually spends £22.50. Again the absolute error is £2.50.

However, it is clear that, in this example, the second estimate is better than the first in the sense that in the first case the error was 25% of the original estimate, while in the second case the error was $12\frac{1}{2}$%. For a comparative measure, relative error is used.

2.6 Relative error

Relative error is the absolute error divided by the estimate. It is often expressed as a percentage:

$$\text{Relative error} = \frac{\text{absolute error}}{\text{estimated figure}} \times 100$$

RELATIVE ERROR

Absolute error = 2.50
Estimate = 10 or 20
Relative error is

$$\frac{2.50}{10} \times 100 = 25\%$$

or

$$\frac{2.50}{20} \times 100 = 12\frac{1}{2}\%$$

2.7 Biased, cumulative or systematic error

If in a series of items the errors are all in one direction, the result will be a biased, cumulative or systematic error.

BIASED ERROR

If people are asked to give their ages at their last birthday, the total age of the group will be lower than the real total:

Actual age		Approximate age
Years	Months	(age last birthday)
15	10	15
18	2	18
17	7	17
18	5	18
70	0	68

The absolute error is 2 years.

The relative error is $\dfrac{2}{68} \times 100 = 3.03\%$.

The aggregate error is normally much larger in biased error than the aggregate unbiased error.

2.8 Unbiased or compensating error

Unbiased error occurs when the approximation is to the nearest whole number or complete unit.

UNBIASED ERROR

If people are asked to give their ages to their nearest birthday the total age of the group is likely to be similar to the real total.

Actual age		Approximate age
Years	Months	(age at nearest birthday)
15	10	16
18	2	18
17	7	18
18	5	18
70	0	70

In this case there is no absolute or relative error because the small amounts added and subtracted balance out so that the final (approximated) total is the same as the true total.

The final total will not always work out exactly as it does in this example, but generally the unbiased error will tend to be small and the greater the number of items the smaller the error will tend to be.

2.9 Calculations involving approximation and error

In both theory and practice it is necessary to make calculations in which approximations and errors are present. This is because so much statistical data is rounded or is accurate to a certain number of significant figures or contains an element of biased or unbiased error. Much of the time the fact that approximation is present in data does not matter, because the degree of accuracy falls well within the levels of tolerance of the material being used and the needs of the subject being studied. However, it is useful to know what influence approximations can have on data and on the results of calculations.

POPULATION STATISTICS

Many population statistics are available only in an approximate or rounded form. Birth rates and death rates are accurate to the nearest whole number. Therefore in making calculations about a town's future population, and commercial development and building based on it, such as the building of schools and the development of cemeteries, the approximate nature of the statistics has to be taken into account.

If it is known that the population of a town is 120 000 to the nearest thousand and the birth rate is 15 births per thousand, accurate to the nearest whole number; it is possible to calculate the limits of error in the statistics:

120 000 to the nearest thousand can be rewritten as $120\,000 \pm 500$.

This can be written in thousands as 120 ± 0.5.

15 to the nearest whole number can be rewritten as 15 ± 0.5.

The greatest possible birth rate would then be:

$120.5 \times 15.5 = 1867.75$ births

The lowest possible birth rate would be:

$119.5 \times 14.5 = 1732.75$ births

All that can be said is that the number of births in the town is likely to be between these two figures:

1800.25 ± 67.5

The difference of 135 births between the two figures could be an important factor in future planning, although for some plans the difference could be considered unimportant.

These examples indicate the need to be aware of the approximate nature of many statistics in making calculations. The basic methods of making calculations of levels of approximations involve addition, subtraction, multiplication and division.

Addition

Example 1. Add 17 ± 0.5 and 3 ± 0.01.

The highest possible result is:

$$17.5 + 3.01 = 20.51$$

The lowest possible result is:

$$16.5 + 2.99 = \underline{19.49}$$

Total   40.00

The mid-point is 20 ($40 \div 2$).
The limits of error (the difference between 20.51 and 19.49 divided by 2) are ± 0.51.

Therefore (17 ± 0.5) + (3 ± 0.01) = 20 ± 0.51.

It can be seen that the error in the aggregate is the sum of the absolute errors in the component parts (0.5 + 0.01).

Example 2. 20 000 correct to the nearest 1000 plus 4700 correct to the nearest 100.

This can be rewritten:

(20 000 ± 500) + (4700 ± 50)

The maximum result is 20 500 + 4750 = 25 250.
The minimum result is 19 500 + 4650 = 24 150.
The mid-point is 24 700 (49 400 ÷ 2).
The limits of error are ±550.
Therefore (20 000 ± 500) + (4700 ± 50) = 24 700 ± 550.

Subtraction

Subtract 20 ± 2 from 100 ± 10.

The maximum result is the highest figure minus the lowest:

110 − 18 = 92

The minimum result is the lowest figure that can be obtained from 100 ± 10 minus the highest figure which can be obtained from 20 ± 2:

90 − 22 = 68

The mid-point is 80 $\left(\dfrac{92 + 68}{2}\right)$.

The limits of error are ±12.

Therefore (100 ± 10) − (20 ± 2) = 80 ± 12.

It can be seen that the error in the answer (12) equals the sum of the errors in the individual parts (2 and 10).

ADDITION AND SUBTRACTION WITH ROUNDED NUMBERS

When adding or subtracting with rounded numbers it is important to remember that the answer cannot be more accurate than the least accurate figure.

Add 327, 631 and 700, where 700 has been rounded to the nearest 100.

327 + 631 + 700 = 1658

But since the least accurate figure is to the nearest 100, the answer must be given to the nearest 100 and therefore the answer will be 1700.

Any attempt to be more exact can only result in spurious accuracy (see Section 2.10).

Multiplication

Multiply 100 ± 2 by 20 ± 1.

The maximum possible result is obtained by multiplying the highest figure that can be obtained from 100 ± 2 by the highest figure that can be obtained from 20 ± 1.

$102 \times 21 = 2142$

The minimum possible result is obtained by multiplying the lowest figure that can be obtained from 100 ± 2 by the lowest possible figure that can be obtained from 20 ± 1.

$98 \times 19 = 1862$

The mid-point is 2002 $\left(\dfrac{2142 + 1862}{2} \right)$.

The limits of error are ± 140.

Therefore $(100 \pm 2) \times (20 \pm 1) = 2002 \pm 140$.

Division

Divide $1000 \pm 2\%$ by $100 \pm 1\%$.

This can be rewritten: divide 1000 ± 20 by 100 ± 1. The maximum result is obtained by dividing the minimum figure that can be obtained from 100 ± 1 into the maximum figure that can be obtained for 1000 ± 20.

$1020 \div 99 = 10.303$

The minimum result is obtained by dividing the maximum figure that can be obtained from 100 ± 1 into the minimum figure that can be obtained from 1000 ± 20.

$980 \div 101 = 9.703$

The mid-point is 10.0003 $\left(\dfrac{10.303 + 9.703}{2} \right)$.

The limits of error are ± 0.3.

Therefore $(1000 \pm 2\%) \div (100 \pm 1\%) = 10.003 \pm 0.3$.

The important factor to remember in these calculations is that the greatest and least approximations are possible and to make allowance for this fact when these figures are used.

In fact, in business, bias may be used purposely in certain circumstances. In estimating future expenditure it may be prudent to base the estimate on biased error to produce a higher figure than an unbiased result.

In estimating individual income it may be useful to use the maximum result of an estimate for some purposes and the minimum result for other purposes. It may be decided that earnings will be about £500 a month (the 'about' meaning say $\pm£40$). This means that earnings could be between £460 and £540 a month or £5520 to £6480 per year, an annual difference of nearly a thousand pounds. For tax purposes the lower figure might be estimated, while in order to obtain a mortgage the higher figure might be used. Adjustments would have to be made when the true figure is known at the end of the year.

2.10   Spurious accuracy

When a figure implies an accuracy greater than it really has, such accuracy can be termed spurious. It is not sufficient to appreciate that complete accuracy is usually impossible, it is also important that claims are not made for such accuracy where it does not exist. For example, to write 6.354 means that an accuracy of up to three decimal places is being claimed. It must mean that because, if the accuracy is only to two decimal places, then the '4' is a guess and it is pointless to include it.

It is easy to be misleading with statistics:

MISLEADING STATISTICS

Example 1
'Buy now and save 100%', may mean that there has been a reduction of 50% from say 10p per unit to 5p. A 100% reduction would mean giving the commodity away free, unless the 100% was based on something else not mentioned (like 100% of a future price increase).

Example 2
To arrive at a high wage it is possible for an employer to add up wages. If one basic working hour at £6 is added to one overtime hour at £9 and one double-time hour at £12, it is possible to arrive at an average hourly wage of £9 (that is £27 divided by 3). In fact, because most of the working hours are basic hours, the average is likely to be nearer £6.

Dr Johnson is supposed to have said: 'Round numbers are always false.' It would perhaps be equally true to say that 'exact numbers are always false'.

EXACT NUMBERS

A commodity is reported to have been sold in the USA for £21 333. In fact the original report may have been that it was sold for about $48 000, this being an approximation of the exact sale price of $48 850. The figure of £21 333 was arrived at by dividing the $48 000 by the day's approximate exchange rate of $2.25 to £1. Therefore an approximate figure has been converted using an approximate conversion rate to produce an 'exact' figure in sterling.

This is spurious accuracy. It would have been more accurate to say that the commodity had sold for about £21 000, or between £21 000 and £22 000.

The statement 'lies, damn lies and statistics' arises from the use of spurious accuracy as well as the careful selection of figures to support a particular argument. Perhaps all statistics need to have a label attached to them: 'treat with care and understanding'.

ASSIGNMENTS

1   Discuss the importance of accuracy in statistics. What is meant by spurious accuracy?

2   Company A produces one commodity. The cost per unit of this commodity is made up of:
   (i)   labour costs of £3 (to the nearest £) per unit,
   (ii)   raw materials costing 90p (to the nearest whole 10p) per unit,
   (iii)   fuel and power costing an average of 10.3p (to three significant figures) per unit,
   (iv)   overhead costs averaging 85.2p (to one place of decimal) per unit.
What is the maximum and minimum cost per unit of output for this commodity?

3   Find examples of rounding in a variety of publications (newspapers, magazines, journals, company reports, government publications). Discuss the degree of rounding in these examples and the results of this rounding on the interpretation of the data.

4   The population of a town is estimated to be 196 000 when the actual population is 200 000. In these figures, what is the:
   (i)   absolute error,
   (ii)   relative error,
   (iii)   percentage error?

5   Discuss the following terms used in statistics:
   (i)   degrees of tolerance,
   (ii)   error,
   (iii)   rounding,
   (iv)   absolute error,
   (v)   relative error,
   (vi)   biased error,
   (vii)   unbiased error.

3 Collecting statistical information

OBJECTIVES

- To describe methods of collecting primary statistical information
- To consider the stages involved in carrying out a survey
- To analyse the processes involved in observation and in interviewing
- To define and describe sampling
- To analyse the basis of sampling
- To describe a variety of sampling methods

3.1 Surveys

Primary data is collected by or on behalf of the person or people who are going to make use of the data. If data is not already available but it is needed to help solve a problem, it has to be collected. This is primary data. **The collection of all primary data involves carrying out a survey or enquiry of one type or another.** A very limited survey or enquiry can be carried out in a few minutes by observation.

A LIMITED SURVEY

How well used is the firm's car park? By looking out of the window it may be observed that there are 50 parking spaces and that five of these are empty. Therefore the car park has 90% usage.

This example is a limited survey given spurious validity by using a percentage. The one rapid observation gives a result only for that particular time of day on that particular day. This could be described as a sample of all possible observations of the car park all day every day. However, it is such a small sample (see Section 3.11) that it is not likely to be very accurate or much help in deciding whether the firm needs a larger car park or could take over part of the car park for some other use.

Two observations would provide a check on each other, and a series of systematic observations would provide a survey with a good chance of providing a useful answer to the question: useful in the sense of helping to make a decision. In the example, argument over whether the car park is fully used or under-used can be narrowed by carrying out a statistical survey. The survey will provide statistical evidence which can help agreement to be reached.

The larger and more detailed a survey the more chance there is of it being both valid and accepted. However, it is better to carry out some type of survey, even if it is very limited, rather than none at all. One rapid observation of the car park is perhaps better than none at all.

THE SMILE TEST

It has been suggested that one way of very quickly testing the morale of staff in offices, factories, schools and colleges is to walk into the building and count the number of people smiling. The more smiles, the higher the morale and job satisfaction.

This 'smile test' should not be taken too seriously! What is required to produce useful results is a systematic survey.

Surveys produce primary data which is collected from basic sources in order to satisfy the purposes of a particular enquiry. Secondary data often provides the framework of information, leaving matters of relative detail to be filled by special surveys.

Survey statistics are available on a wide range of topics.

- *Government surveys* form an important part of the total available.
- *Market research surveys* are in the main carried out for a particular client and the results are usually produced in a private report.
- *Research surveys* are carried out by academics and other research workers often with the results published in journals.
- *Trade associations* collect statistics on sales for their members to help establish their share of the market.
- *Firms* commission *ad hoc* surveys on a wide variety of subjects.

Apart from the limitations of time and costs, there are few limits to the nature or quality of data that can be obtained by means of a survey.

3.2   Survey methods

Whether a survey is a 100% survey of all possible items or is a small sample survey, there will be a series of stages in carrying it out.

The survey design

This depends on the subject of the survey, what methods are available and the amount of time and money that can be spent collecting information. The UK Census of Population is unusual, because it is carried out on a 100% basis, aiming to find out a very wide range of information from every household in the country. On the other hand the *Family Expenditure Survey* is concerned mainly with households' income and expenditure and it is based on a sample. The *FES* employs a variety of methods so that the design of the survey is fairly complex.

The survey design includes decisions on the size of the sample and the type of sample to use as well as decisions on the method of collecting the information.

The pilot survey

This is a preliminary survey carried out on a very small scale to make sure that the design and methodology of the survey are likely to produce the information required. The survey may be tried out on two or three people instead of 2000 or 3000. It is then possible to alter the design of the survey if it is discovered to be inadequate, before time and money have been spent on the main survey.

The collection of information

The main methods of collecting primary data are through observation, interview and questionnaire. Most surveys use a combination of these methods (see Sections 3.3, 3.4, 3.5).

Coding

It is useful in processing survey forms to pre-code the questions so that answers can easily be classified and tabulated. Coding can simply mean numbering or lettering questions or can include more elaborate methods of identifying groups of answers.

Tabulation

Once the information has been collected it has to be classified and tabulated (see Sections 4.2 and 4.3). In designing a survey it is useful to consider the problems that might arise at the tabulation stage.

Secondary statistics

The information contained in the tables will often need to be summarised by calculating secondary statistics such as percentages and averages.

Reports

The final stage of a survey is usually to write a report on the results and to illustrate the results with graphs and diagrams (see Section 4.5 and Chapter 5).

There are a number of methods of collecting primary data: observation, interviews and questionnaires. Questionnaires are often used with interviews, particularly formal interviews, and with systematic observation. Whichever method or
combination of methods is used the important factor is to obtain accurate information. Although the methods are often used together, it is easier to consider
them separately here.

3.3 Observation

Direct observation

**Direct observation can be used to discover a variety of types of information
including aspects of social and economic behaviour.** It has been used to look at
consumer behaviour, working methods and a range of social activities.

Direct observation is the classic method of scientific enquiry; biologists,
physicists, astronomers and other natural scientists rely on centuries of systematic observation for their accumulated knowledge. In the social sciences
observation can be used as a method of watching humans, as if observing
animals, in a detached, relatively objective way. Other methods of collecting
primary data tend to be less objective.

OBSERVATION

Observation is the method used by consumers to compare prices, and by
companies to gather information on the use of their car park or to inspect the
quality of their products.

In many cases the observer will try to be as unobtrusive as possible in order
not to participate directly in the events being observed. This may be difficult
unless the observer is hidden. In time and motion studies the investigator has
often been caricatured as someone with a stop-watch, notebook and binoculars,
trying to watch people working without being seen. The implication is that if the
investigator was seen the workers might work harder than usual. In the same
way an investigator sitting in a classroom to observe the behaviour of the children or the teacher could well influence their behaviour. The children might
behave better than usual because they want to impress the stranger; or they
might behave worse because they want to show off.

In an attempt to avoid these problems there have been approaches on two
extremes: participant and systematic observation.

Participant observation

In participant observation the observer becomes a participant in the activity
being observed. The observer works in the factory to observe how hard people
work, or takes part in classroom activity to observe the behaviour of children

and teachers. This approach is very time consuming, and the observer does not know the extent of his or her own influence on what is happening.

Systematic observation (or objective observation)

Systematic observation is used at the other extreme to observe only events which can be investigated without the participants knowing. An example is road use. Observers can watch a section of road and note the number and type of vehicles at various times of day on different days. This method is very objective, but the motives of the people observed are not questioned. It is not known why the drivers are travelling on that particular road at that particular time. To overcome this problem, drivers can be asked the purpose of their journey. This may still be systematic but will not be objective in the sense that the participants will now know what is happening.

It is open to question how far people should be observed without their knowledge and at what point this becomes an invasion of privacy. Observation has been applied to methods of changing people's behaviour and it is a matter of opinion how far this should go.

EXAMPLE

It was hoped that background music in supermarkets would increase the level of activity and either encourage people to buy more or to shop faster. Observation showed that when loud music was played rather than soft music, sales did not change but people did shop faster.

Problems of direct observation

- *Objectivity* – to remain objective, observers cannot ask the questions which will help them to understand the events they are observing.
- *Selectivity* – observers can be unintentionally selective in perception, recording or reporting.
- *Interpretation* – observers may impute meanings to the behaviour of people which the people themselves do not intend.
- *Chance* – a chance event may be mistaken for a recurrent one.
- *Participation* – observers can influence events because people realise that they are being observed and change their behaviour.

Mechanical observation

Mechanical observation is used in survey methods under certain conditions. For example, the number of vehicles passing a particular point on the road can be recorded mechanically (although it is difficult for mechanical means to distinguish between types of vehicle). Meters are used to measure television audiences by recording the length of time the set is switched on and the channel to which it is tuned. Mechanical methods of inspection and testing quality are commonplace in industry (see Section 8.8).

Sophisticated mechanical means (such as television, film and tape recorders) are used to provide more complex information. The disadvantage of these methods is that they can be expensive and, if over used, can provide the survey team with too much unselected information.

Mechanical means of observation do enable much more detailed material to be collected than would be possible by an observer working alone. Since the information is recorded it can be analysed in depth some time later. Also, using a piece of apparatus may avoid the influence of the observer on events (although people being observed may still be influenced if they know they are being checked).

3.4 Interviews

An interview can be described as a conversation with a purpose.

In an informal sense everybody uses interviewing to obtain information. 'What was the score in the football match last night?' If the answer consists only of the score and the questioner wants to know more about the game, he or she may have to ask more questions. 'Who scored?', 'When were the goals scored?', 'How did player X play?', 'Was there a large crowd?' and so on.

In many cases questioning will start with general questions, or with questions that are easy to answer. More specific and more complex questions can follow up certain points to clarify them and provide a greater understanding of the subject.

Similar methods are used in more formal interviewing. **A 'formal' interview is a conversation between two people that is initiated by the interviewer in order to obtain information.** An interview is likely to be more structured than ordinary conversation, because the interviewer will present each topic by means of specific questions and will decide when the conversation on a topic has satisfied the objectives of the interview.

Interviews are used in a wide variety of circumstances and for many purposes.

- *Attitudes:* opinion polls are based on interviews to discover people's attitudes towards a proposed product that a company is developing, or to test people's political views, or to measure the change made to people's views or actions by particular events (such as an advertising campaign).
- *Motives:* interviews are used in an attempt to discover why people are behaving in a particular way. Road usage surveys include interviewing drivers to ask them where they are going and why they are making their journey.
- *Job selection:* more or less formal interviews are used for selecting employees.
- *Reporting:* interviewing has reached a sophisticated level on radio and television.

The great advantage of interviewing over direct observation is that it is possible to question a person's motives and attitudes, about changes in behaviour and about possible future behaviour.

Interviews may be formal or informal.

In **formal interviews** set questions are asked. Surveys on road use tend to be formal, with all drivers being asked the same list of questions.

In **informal interviews** the questions may follow a pattern but will vary in order and content between interviews. Job interviews are often relatively informal with similar questions being put to each candidate, but varying in response to the answers received. Some employers use two interviews for their recruitment, one very informal, the other formal.

Interviews are not uniformly successful in obtaining the information required. Respondents differ in ability and motivation, interviewers differ in skill and experience, and the interview content differs in feasibility. Interviews are an interactive and subjective method of finding information. Formal interviewing attempts to reduce these influences. This can be carried to the extreme of erecting a screen between the interviewer and the respondent. Usually the two are face-to-face and they may take an instant liking or disliking to each other.

The way questions are asked may influence the answers. For example, it is not unusual for electoral canvassing returns for the same area of a constituency to show that both major parties have a majority of the voters pledged to vote for them. This 'impossible' situation can arise because a number of voters may want to give the answer most likely to please the canvasser, or may feel that by giving an expected answer the interview will be over quickly. This type of interview is often carried out in the evening, on the doorstep of the voter's house, with a favourite television programme on in the background. All of these factors influence the success of the interview measured in terms of the accuracy of the information received.

The success of an interview will depend very much on the two protagonists involved, the interviewer and the respondent.

The respondent

The ability of the respondent to 'make a success' of the interview will depend on the following.

- *The accessibility of information* – the information being sought by the interviewer has to be accessible to the respondent so that he has it clearly thought out and is able to express it in the terms used by the interviewer.
- *Role* – the respondent needs to be clear about his role particularly in the sense of knowing what information is relevant and how completely he should answer.
- *Motivation* – the respondent needs to be motivated to answer questions and to answer accurately. The interviewer can suggest why it is in the respondent's interests to answer the questions. Up to a point people are happy to give interviews because they are asked; for long interviews greater incentives may be required. People will often answer questions at length if they believe that they can influence events (such as the siting of a new airport, or the route of a new motorway). Payment may help to provide an

incentive, but can encourage people to become respondents even when they know little about the subject.

- *Prestige* – if the interview is carried out for a well-known or prestigious company or institution, or has government backing, respondents may be encouraged to answer questions carefully.

The interviewer

The interviewer may influence the results of the interview in obvious ways, such as careless recording of answers, or poor reporting of results, or by cheating. It has been known for interviewers to fill in questionnaires themselves to save themselves the trouble of collecting the answers. There are more subtle ways in which the interviewer may influence the result; these are referred to as *interviewer bias*.

Interviewer bias may result from:

- the way the questions are asked;
- the extent of probing or asking supplementary questions and the kind of answers the interviewer expects;
- the interviewer expecting that early replies of a respondent establish a pattern which will be followed in later answers;
- the interviewer expecting a person of a certain age and appearance to answer in a particular way.

Added to the problems arising from respondent expectations and interviewer bias are problems common to any questionnaire (see Section 3.5). The organisation and sequence of questions may influence the results of the interview.

All these problems can be reduced by good selection, training and supervision of interviewers so that they are aware of the problems that can arise and can try to avoid them. Market research organisations use experienced interviewers for their door-to-door and street-corner interviews.

Used carefully, the interview is an excellent method of collecting quite complicated information. In practice the amount of information collected by interviewing is limited by time and cost.

3.5 Questionnaires

A questionnaire is a list of questions aimed at discovering particular information.

Questionnaires can be distributed by hand (pushed through people's letter-boxes) or by post. Frequently they are used in interviews to provide the interviewer with a set list of questions to ask.

Advantages of postal questionnaires

- They are relatively cheap to distribute.
- Therefore they can be sent to large numbers of people.
- The answers can be carefully considered.

Disadvantages of postal questionnaires

- It is not possible to explain questions or to follow them up.
- A poor response is usual, unless there is a very strong incentive to return the questionnaire.

Questionnaires distributed by hand usually have a more limited distribution because of the costs involved. However, the response rate may be much higher because they can also be collected by hand, and this provides an incentive to complete the form.

In all questionnaires problems can arise from the design of the questions and the design of the questionnaire form. There are a number of points that need to be considered.

FACTORS INVOLVED IN QUESTION DESIGN

- Questions should be simply and clearly worded so that they can be understood by the 'average' respondent.
- Questions must be clearly useful and relevant, and designed to produce the desired information.
- Questions must be free from bias.
- Questions must not be so personal and private that the respondent will be reluctant to give an honest answer.
- The order of questions needs to be logical and to help the respondent to remember the answers.
- Questions should be unambiguous and should not include vague words, such as 'fairly' and 'generally'.
- Leading questions need to be avoided, for example: 'don't you think something should be done about . . . ?', encourages a positive answer.
- Hypothetical questions are of limited value.

FACTORS INVOLVED IN THE DESIGN OF THE QUESTIONNAIRE FORM

- It needs to be clear from the form who should complete it.
- It needs to be clear where the answers should be recorded.
- Sufficient space should be available to complete the answers.
- The questions need to be set out so that it is clear how they follow on.
- As many questions as possible should be able to be answered by 'yes' or 'no', or by the respondent deleting a word or phrase, or by ticking a box.
- The convenience of both the respondent and the survey team who will process the questionnaires need to be considered.

Everybody fills in forms of one kind or another, for the Inland Revenue, for VAT, to enrol on a course, to claim social benefit, to register a car, and so on. Many forms are poorly designed (inappropriate space for replies, for example), or include ambiguous questions. This indicates the difficulty of producing a well-designed form and clearly understood questions. The objective of producing a questionnaire is to collect information, therefore it is worth organising and designing it carefully.

3.6 Why sample?

A sample is anything less than a full survey of a 'population'; it is usually thought of as a small part of the population, taken to give an idea of the quality of the whole. **The 'population' is the group of people or items about which information is being collected.**

It may seem desirable to base decisions on complete counts or measurements of people and commodities. Anything less than this may be felt to include only part of the information and to be open to a high degree of error and approximation. In practice it is only in limited circumstances that a 100% survey can be completed. The UK Census of Population every ten years is exceptional because it is a 100% survey. Within a company it may be possible to carry out a comprehensive survey of the labour force on a variety of subjects.

However, for much general information about people's opinions and attitudes and about the quality of commodities it is often impossible to carry out complete surveys. A marketing manager might like to carry out a census of all women on their attitudes to a product; a works manager might like to inspect in detail every item coming off the production line. In neither case is this feasible; it would be impossible for every marketing decision or every aspect of controlling quality to be based on a complete census of millions of people or thousands of items. A complete survey is not only impracticable but also unnecessary. In many cases a sample is preferable.

3.7 Advantages of sampling

Cost

It is cheaper to collect information from 2000 people than from two million. However, the cost per unit (or item, or person) may be higher with a sample than with a complete survey. More skilled personnel may be used, new costs may be added such as those involved in sample selection and in calculations of the precision of the sample results. Also, overhead costs are spread over a smaller number of units. In spite of these extra costs, samples are usually so much smaller than a complete coverage (often 10% or less) that total costs are likely to be very much less.

Time

Information is often required within a specified time, so that a decision can be made and action taken. A sample requires less fieldwork, tabulation and data processing than a full survey. Also, following up non-response and other problems is quicker with a sample than a full survey because there are fewer items.

Reliability

A high level of reliability can be achieved because fewer units are surveyed in a sample than in a full survey and therefore resources can be concentrated on obtaining reliable information. Well-trained field staff can be employed, more checks and tests made and more care taken with editing and analysis. Respondents may be more willing to provide detailed information if they know that they form a small sample of the population. They may feel that because they are representing the population they should provide reliable information.

In fact absolute accuracy may not be required. The size of a sample can be adjusted so that the resulting accuracy is sufficient for a decision to be made (see Section 8.5). If a larger sample is taken, or a full survey, resources are being wasted.

Resource allocation

By using samples it is possible to carry out several studies concurrently, and therefore use resources efficiently.

Product destruction

Some sample tests destroy the product (examples include light bulbs and television tubes). No products would be left unless samples were used.

This last point illustrates the fact that **the interest is not in the sample items except in so far as they may be used to draw inferences about the population from which they are selected.**

IMPORTANCE OF SAMPLES

A market research team will want to draw inferences about 20 million women, not about the 3000 actually interviewed; the quality control inspector is interested in the thousands of components produced each day, not the few that he has tested to destruction.

3.8 Objectives of sampling

Descriptive statistics

The most common objective of sample surveys is **to estimate certain population statistics or parameters** (such as averages and proportions).

A sample is selected, the relevant statistic is calculated and this is used as an estimate of the population statistic. The statistic should be accompanied by a statement about the accuracy of the result in terms of standard error. Standard error is a method of indicating the variability of a sample statistic; for instance to show the extent to which a sample average deviates from the population average (see Section 8.5).

STANDARD ERROR

The average hours of overtime worked per week by the employees of a large company is estimated by carrying out a 10% sample survey. The average overtime is calculated (say 4.5 hours per week), with a statement about the standard error involved (say 0.5 hours). This indicates that the sample average may deviate by half an hour from the population average.

Statistics of inference

Another use of sampling is **to test a statistical theory about a population.** A theory or hypothesis is held about a population, a sample survey is carried out and the results are then interpreted to test whether the results support or refute the theory (see Section 8.6). For example, it may be thought that the average hours per week of overtime worked by the employees of a large company is five hours. A sample survey is taken to test this theory.

The underlying objective of sampling is to describe the population from which the sample is taken. If sample results are to be used for decision making it is very important to assess the reliability of these results. It would be unfortunate for a company to reorganise on the basis of a 10% sample if the particular employees questioned in fact represent a distinct minority of all the employees in the company.

3.9 The basis of sampling

The possibility of reaching valid conclusions concerning a population from a sample is based on two general laws: the law of statistical regularity and the law of the inertia of large numbers.

The law of statistical regularity

This states that a reasonably large sample selected at random from a large population will be, on average, representative of the characteristics of the population.

For this law to work, the selection of the sample must be made at random, so that every item in the population has an equal chance of being included in the sample. Also, the number of items in the sample must be large enough to represent the whole population and to avoid undue influence of extreme items on the average. The larger the number of items selected, the more reliable the information will tend to be. For instance, if three people are asked which political party they support there could be equal representation for three parties or a 'minority' party might be supported by all three people, even if it is known that most people support one or other of the two major parties. A sample of 300 people would be likely to provide a better impression of the support for each party and a sample of 3000 people might represent the views of the population quite well.

The law of the inertia of large numbers

This states that large groups of data show a higher degree of stability than small ones. There is a tendency for variations in the data to be cancelled out by each other. If a large number of items is sampled, it is unlikely that the variations in them will all move in the same direction. If, for example, the length of nails coming off a production line is supposed to be 2.5 centimetres, some of the nails will be a fraction longer than this and some a fraction shorter, so that on balance the average length will be 2.5.

> The law of statistical regularity and the law of the inertia of large numbers are part of a mathematical theorem, the central limit theorem, and are ways of describing the theory of probability. The theory of probability is the basis of statistical induction, which is the process of drawing general conclusions from a study of representative cases (for a further discussion of this theory and the basis of sampling see Chapter 8).

3.10 Sampling errors

Errors in a sample survey may arise both from sampling errors and from non-sampling errors or bias.

- **Non-sampling error: this is due to problems involved with the sample design.** Many non-sampling errors would arise with a full survey, but some of them are due specifically to sample design, including such factors as the choice of a sampling frame and sampling units (Section 3.12 below).

- **Sampling error: this is the difference between the estimate of a value obtained from a sample and the actual value.** A sample may show that the average weekly wage of a group of employees is £100, when the actual average is £110 a week. The sampling error is £10.

Sampling errors arise because, even when a sample is chosen in the correct way (by random methods), it cannot be exactly representative of the population from which it is chosen. The degree of sampling error will depend on the size of the sample; the larger the sample the smaller the error (see Sections 3.12 and 8.5). This is not dependent on the size of the population. A population of 3 million does not require a larger sample than a population of 300 000 or 30 000.

The important point about sampling error is that, provided the sampling method used is based on random selection, it is possible to measure the probability of errors of any given size. The total error in a sample arises from both non-sampling and sampling errors and cannot be substantially reduced unless both types of error are simultaneously controlled. There is no point in taking a larger sample in order to reduce the sampling error if there are design faults in the sample.

3.11 Sample size

The size of a sample (the number of people or units sampled) **is independent of the population size.** It does depend on the resources available and the degree of accuracy required. Other things being equal a large sample will be more reliable than a small sample taken from the same population.

Therefore the number of items sampled is a matter of judgement based on the variability in the population. A population which is known to be very variable (including numbers of people with different opinions or including units of many types) will require a larger sample to represent it than a population known to be very homogeneous.

SAMPLE SIZE

Samples of political opinions in the UK have increased in size in recent years because it is felt that the electorate has become more variable and volatile in its opinions, and a larger sample is more likely to be representative of the whole electorate.

3.12 Sample design

The sampling population is the group of people, items or units under investigation.

The sample units are the people or items which are to be sampled. These units need to be defined clearly in terms of particular characteristics. For instance, in a sample of 'vehicles', this term needs to be defined in terms of say motor cars, buses and commercial vehicles. This would exclude motor cycles, bicycles, tractors and other vehicles that use the road. In practice it may be found that motor cars, buses and commercial vehicles need to be defined very clearly in their turn.

The sample frame is the list of people, items or units from which the sample is taken. The sample unit is defined and then a suitable sampling frame is sought. It should be comprehensive, complete and up-to-date to keep bias to a minimum. General examples of sampling frames include electoral registers, telephone directories, wage lists.

The survey method includes designing questionnaires and deciding how to distribute them or how to carry out interviews or observations.

Sampling methods fall into two categories: **random samples** (simple, systematic, stratified); **non-random samples** (multi-stage, quota, cluster). The type of sampling method chosen depends on the nature and purpose of the enquiry. It is difficult to assess the sampling error involved in non-random samples and therefore in using these methods there is usually an attempt to include some element of randomness.

Principles of sample design
- To avoid bias in the selection procedure
- To achieve the maximum precision for a given outlay of money and time

How to design a sample
- Decide on the objectives of the survey.
- Assess the resources available.
- Define the sample population and the sample unit.
- Select a sample frame.
- Decide on a survey method.
- Choose a sampling method.

3.13 Bias in sampling

Bias consists of non-sampling errors, which are not eliminated or reduced by an increase in sample size. Bias may arise from the following.

- **The sampling frame:** if this does not cover the population adequately or accurately.
- **Non-response:** if some sections of the population are impossible to find or refuse to co-operate. There are certain groups that tend to be underrepresented in many surveys; these groups include the very rich and very poor, young adults, working women.
- **The sample:** if the most 'convenient' sample is selected it may be biased and non-random. Examples are shoppers in a particular shopping centre at a particular time of day, or the top box of components in a delivery.

- **Question wording:** poorly worded or ambiguous questions and interviewer bias may cause problems in sample surveys in much the same way as they do in comprehensive surveys.
- **The sample unit:** a personal element may enter into selection. Substituting one unit or person for another may introduce bias.

Any of these factors may cause systematic, non-compensating errors in a sample survey.

BIAS IN SAMPLING

In 1936 in the USA the *Literary Digest* carried out a huge sample of 10 million individuals, yet its forecast of the result of the US presidential election was wrong, because:

- the sample was picked from telephone directories which did not adequately cover the poorer section of the electorate;
- only 20% of the mail ballots were returned and these probably came predominantly from more educated sections of the population.

The sample returns indicated that Franklin D. Roosevelt would be defeated, whereas in fact he was elected with one of the largest majorities ever recorded in the USA.

3.14 Sampling methods

There is a range of sampling methods to choose from in carrying out a sample survey.

Simple random sampling

In random sampling each unit of the population has the same chance as any other unit of being included in the sample.

Random numbers

To select each unit on a random basis a lottery method can be used. For large groups it is not possible to number or name every unit of the population and then pick them out of a hat. Therefore random numbers are used, such as the system for selecting British Premium Bonds through the Electronic Random Number Indicating Equipment: ERNIE.

Sampling with and without replacement

Unrestricted random sampling is carried out 'with replacement'. This means that the unit selected at each draw is replaced into the population before the next draw is made, so that a unit can appear more than once in the sample.

In sampling without replacement only those units not previously selected are eligible for the next draw.

In applied statistics it is assumed that sampling is without replacement (in a lottery a winning ticket is not usually replaced in the hat or box to allow it to win a second prize.) **Simple random sampling is sampling without replacement.**

Main features of simple random sampling

- It is 'simple' compared with more complex methods which are also random (such as systematic and stratified sampling). In the USA it tends to be referred to as 'simple' probability sampling.
- The technique of randomisation ensures the validity of the techniques of inference such as deciding the confidence with which results can be accepted (see Sections 8.6 and 8.7).
- Simple random sampling is the standard against which other methods are evaluated.
- It is suitable where the population is relatively small and where the sampling frame is complete.

Systematic sampling

This is a form of random sampling, involving a system. The system is one of regularity. The sampling frame is taken and a name or unit is chosen at random. Then from this chosen name or unit every nth item is selected throughout the list.

SYSTEMATIC SAMPLING

If the sampling frame contains 100 000 names and a 2% sample is required, the 2000 names can be selected at regular intervals. The first is selected at random from the first 50 names (50 because $50 \times 2000 = 100\,000$). If the 35th name is picked the names are selected at regular intervals to make up 2000 in all (the 35th, 85th, 135th, 185th . . .).

Random route sampling

This is a form of systematic sampling used in market research surveys. It is used mainly for sampling households, shops, garages and other premises in urban areas.

An address is selected at random from a sampling frame (usually the electoral register), as a starting-point. The interviewer is then given instructions to identify further addresses by taking alternate left- and right-hand turns at road junctions and calling at every nth address (shop, garage, etc.) *en route.*

Stratified random sampling

This is a form of random sampling in which all the people or items in the sampling frame are divided into groups or categories which are mutually exclusive (that is, a person or unit can be in one group only). These groups are called 'strata'.

Within each of these strata a simple random sample or a systematic sample is selected. The results of the sample for each stratum are processed. If the same proportion (say 5%) of each stratum is taken, then each stratum will be represented in the correct proportion in the overall result. This eliminates differences between strata from the sampling error.

STRATIFIED RANDOM SAMPLING

In a marketing survey the sales of cigarettes in a variety of outlets may be investigated by dividing the retail outlets into strata. In a particular town or urban area shops may be divided into large, medium and small outlets and a simple random sample taken based on shops from each category. A clear definition of the strata is important so that there is no overlap between shops.

Advantages of stratified random sampling
- It may provide a more accurate impression of the population (where there are clear strata) than other sampling methods.
- This sampling design may be an improvement on a simple random sample for certain populations.

Disadvantages of stratified random sampling
- If the strata cannot be clearly defined, the strata may overlap, reducing the accuracy of the results.
- Within the strata, the problems are the same as for any simple random sample or systematic sample.

The characteristic of random, systematic and stratified sampling is that every individual has a known probability of being included in the sample. Therefore these methods can be called random or probability sampling.

Non-random or judgement sampling methods are used when random methods are not feasible. This may be:

1 when searching for the people selected by random methods would be a long and uneconomic task;
2 when all the items in the population are not known and either there is no suitable sampling frame or it is incomplete;
3 when a random sample would involve expensive travelling to find respondents to interview.

These three categories coincide with the three most common non-random methods of sampling: quota, cluster and multi-stage.

Quota sampling

In quota sampling the interviewer is instructed to interview a certain number of people with specific characteristics.

The quotas are chosen so that the overall sample will reflect accurately the known population characteristics in a number of respects. Quota sampling can be described as non-random but representative stratified sampling.

In the example on page 45 the table highlights the fact that the more characteristics that are introduced the more difficult the interviewer's task becomes.

QUOTA SAMPLING

Interviewers may be told to interview, over a period of several days, 50 people divided into age and socio-economic groups, to ask them their opinions on a television advertisement. These groups may be divided in proportion to the numbers in the population. Therefore the instructions to the interviewers could be to interview on the basis of the following table.

Interview selection table for quota sampling

Age groups	Socio-economic groups	Numbers	
16–25	A/B	1	
	C	3	
	D/E	1	5
25–45	A/B	3	
	C	9	
	D/E	3	15
45–65	A/B	4	
	C	10	
	D/E	6	20
65 and over	A/B	2	
	C	6	
	D/E	2	10
	Total		50

Already finding the individuals of the right age and socio-economic groups in the numbers indicated on the table may be difficult. If a further division was made, for instance into male and female, each of these numbers would have to be further subdivided.

In this example, the instructions to the interviewer may be to go to a particular shopping area at a certain time to interview people in the numbers indicated on the table. The first people encountered who fit the characteristics listed are interviewed.

It can be argued that the greatest defects in sampling are at the interview stage, and in processing the data and therefore that the sample itself is a small source of error; therefore it can be argued that the disadvantages of quota sampling are outweighed by the advantages. The fact remains that unless a random element is introduced in the selection of the sample it is not possible to estimate the sampling errors.

Cluster sampling

In cluster sampling (or area sampling) clusters are formed by breaking down the area to be surveyed into smaller areas; a few of these areas are then selected by random methods, and units (such as individuals or households) are interviewed in these selected areas. The units are selected by random methods.

Cluster sampling is popular where the population is widely dispersed and it is easier to sample a cluster of people than a range of people or households over a wide area.

It is often used to survey the distribution and possible markets for consumer durables such as television sets and washing machines. Also it is used for quality control where batches of items are removed from the production line for testing and inspection.

Multi-stage sampling

Multi-stage sampling is a series of samples taken at successive stages.

1 The country may be divided into geographical regions.
2 A limited number of towns and rural areas are selected in each region.
3 A sample is taken of people or households in the selected towns and rural areas.

CLUSTER SAMPLING

A map of an urban area is divided by a grid and a selection of these areas is taken at random (see Figure 3.1).

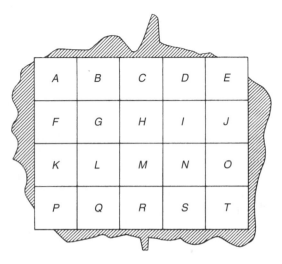

Figure 3.1 Urban area grid for cluster sampling

Perhaps areas *D*, *G* and *J* are chosen. Every household in these areas can be interviewed, or systematic or random route sampling used. A team of interviewers can be sent to the selected areas so that the survey can be carried out quickly.

Advantages of cluster sampling
- Where there is no suitable sampling frame this may be the only possible method.
- Time and money are saved in travelling between locations and searching out respondents, because interviews are concentrated in a few small areas.

Disadvantages of cluster sampling
- Clusters may comprise people with similar characteristics (areas *D*, *G* and *J* may all be relatively wealthy areas) and therefore the results may be biased (this can be reduced by taking a large number of small samples).
- Although there are elements of random sampling in this method it is often difficult to estimate sampling errors.

Often the selection of towns and rural areas is carried out in such a way that the probability of selection is proportional to the size of the population. Therefore a town with a population of 120 000 would stand ten times the chance of being selected as a town with a population of 12 000. Correspondingly, an individual in the larger town stands one-tenth the chance of being selected compared with an individual in the smaller town, so that individuals in both towns stand an equal chance of selection at the beginning of the sampling operations.

This may be the only practical method of sampling when the population of the whole country is being surveyed. Other methods would be too time-consuming and expensive. Regional differences can be allowed for by selecting regions at the first stage. At the second stage a random sample may be used, although again the proportion of urban against rural areas may be allowed for by a proportional distribution. At the third stage simple random or systematic sampling may be used to arrive at the individuals or households to be surveyed.

MULTI-STAGE SAMPLING

The *Family Expenditure Survey* presents an excellent example of the multi-stage sampling method. It is a continuous survey introduced in the UK by the government in 1957 and conducted by the Office of Population, Censuses and Surveys. Each year a sample of over 10 000 households is selected.

1   The country is divided into 1800 areas, from which a stratified sample of 168 areas is chosen. Stratification is by region, type of area (urban/rural) and by house value. This is to ensure that the areas selected are representative of the whole country.
2   Selected areas are divided into districts and four districts from each are selected by systematic sampling.
3   From each chosen district 16 households are selected by taking a systematic sample from the electoral register. The total number of households selected is therefore 16 (households) × 4 (districts) × 168 (areas) = 10 752.

Of these 10 752 households about three-quarters co-operate. These are visited and each member of the household over 16 is asked to keep records covering a period of two weeks of all payments made. In addition, the interviews complete questionnaires on each household covering items of regular expenditure such as gas and electricity bills, details of occupation, income, age and marital status.

All the documents are checked by the survey officials and then forwarded to the Department of Education and Employment for analysis and publication. The results of this survey form the basis of the Index of Retail Prices (see Section 9.5).

There is a range of other sampling methods used under particular circumstances. These include multi-phase sampling, replicated sampling, master samples and panels.

Multi-phase sampling

This is a type of sample design in which some information is collected from the whole sample and additional information is collected from sub-samples of the full sample, either at the same time or later.

In a survey of households basic data may be needed from all the households. This may include the occupations of the members of the household, household income, ages and so on. However some data may be required from only a small sub-sample because:

- the data is less important;
- some factors may be known to be constant in the population and therefore reasonable accuracy can be achieved from a smaller sample;
- some information may be so costly or troublesome to collect that it is only possible to survey a small sample.

This method of sampling can reduce costs and reduce the burden of work on respondents. It was used in the UK Census of Population in 1971 to collect more detailed information on people's qualifications in science and technology.

Replicated or interpenetrating sampling

This is when a number of sub-samples, rather than one full sample, are selected from a population. All the sub-samples have exactly the same design and each is a self-contained sample of the population.

REPLICATED SAMPLING

A full sample of 500 employees could be divided into two sub-samples of 250 employees, or five sub-samples of 100 employees and so on.

Advantages of replicated sampling
- If the size of the total sample is too large to permit the survey results to be ready when they are wanted, one or more of the replications can be used to obtain results in advance.
- These samples can throw light on variable, non-sampling errors, because each of the sub-samples produces an independent estimate of the population characteristics. Therefore, if each sample is carried out by a different interviewer, it is possible to obtain an estimate of variation between interviewers.

Disadvantages of replicated sampling
- Problems can arise from the cost in time and money of carrying out a series of samples.

Master samples

These are samples covering the whole of a country to form the basis (that is, to provide a sampling frame) for smaller, local samples.

The US government has carried out master samples of agriculture to provide a framework for local agricultural surveys.

The units in the master sample must be fairly permanent or long lasting if the results are to be useful for any length of time.

Panels

A group of people is selected from a survey population by a random sample. They form a panel of people who are surveyed at various times over a period of time. This means that the same information is asked for from the same sample at different times.

PANELS

The attitudes of a panel of car owners can be surveyed before and after an advertising campaign. The results can be compared to see the influence of the advertising campaign.

Advantages of panels
- Changes and trends in behaviour and attitudes can be surveyed.
- The effects of specifically introduced measures can be estimated.
- The survey team can study those who change their views and those who do not.
- Evidence may be produced on the ordering of variables – which was cause and which was effect. An ordinary survey can show an association between a worker's attitude to his/her job and his/her position in the firm, but it does not indicate which came first.
- Overhead costs can be spread over a period of time.

Disadvantages of panels
- It may be difficult to recruit the initial sample: people may not want to be involved with a panel.
- Panel mortality tends to be high – people leave panels for a variety of personal and other reasons, there is often a 50% drop-out over a few weeks.
- Panel conditioning means that the members may become untypical of the population they represent – television panels start to view different programmes or concentrate more on programmes because they know that they will have to answer questions on the programmes.

The sampling discussed in this chapter should not be confused with conscious or purposive sampling. This consists of taking carefully chosen samples to present the 'best' units, not typical or representative units.

ASSIGNMENTS

1   Select any form or questionnaire and analyse the good and bad points of its design.

2   Design a questionnaire for a company which is considering the development of its social and sporting facilities. The questionnaire is to be distributed to all the employees of the company in order to discover their attitudes to these developments.

3   Carry out a survey on the road usage of a stretch of road, or the use of a car park. Analyse the results of the survey and write a short report on these results and on the problems of carrying out the survey.

4   Interview five people to discover their attitudes towards a four-day working week for:
    (i) other employees,
    (ii) themselves.

5   Discuss the importance of the concept of randomness in sampling. If it is so important why are sampling methods used which do not include random methods?

6   Find a description of a sample survey and summarise in no more than 500 words the main sampling method used.

7   Design a sample survey to discover the likely success of a new consumer durable, and carry out a pilot survey (consider possible products and names).

8   What are the objectives of sampling? Consider why sample surveys are carried out in preference to full surveys.

9   Carry out a series of dice-throwing exercises consisting of a total of 100 throws. Record the numbers that come up on each throw and the numbers that come up in each set of 10 throws, in the form of a table. Draw a curve to illustrate the table. Write a report to indicate the effects of changing sample size and to compare the results against the expected frequencies.

4 How to present statistics (1)

OBJECTIVES

- To describe the main methods of presenting data
- To provide an understanding of the construction and use of tables
- To provide an understanding of the construction and use of frequency distributions, histograms and frequency curves

4.1 The aims of presentation

Primary and secondary data need to be arranged and presented in some way before the information contained in the data can be interpreted. Primary data may be in the form of a pile of completed questionnaires or a long list of figures; secondary data may be contained in government publications, company reports, books and archives from which relevant information needs to be collated.

The aim of presenting figures is to communicate information. Therefore the type of presentation will depend on the requirements and interests of the people receiving the information.

PRESENTATION

To compare different methods of presenting the same information, all the national newspapers can be bought on the day after a Budget. It is then possible to compare the ways in which the figures contained in the Budget are presented. The broadsheet papers tend to print tables of figures and an almost verbatim report of the Budget. The tabloids publish diagrams, drawings and brief tables, with a short summary of the Budget speech in Parliament.

Figure 4.1 *Sales graph for company A*

When data is presented it is important to provide information clearly and at the same time make an impact. A sales graph which shows a sharp decline in sales from the date a new sales manager arrived is a good example (Figure 4.1).

It has been suggested that the facts do not matter as much as the way they are presented. Obviously facts do matter, but if they are poorly presented they may be overlooked or misinterpreted. Therefore the way in which statistical data is presented is important.

The form of the presentation should be based on the following factors:

- **clear presentation** of the subject matter;
- **clarification** of the most important points in the data;
- the **purpose** of the presentation;
- the amount of **detail and accuracy** required;
- the most **appropriate method** of presentation.

For example, presenting sales figures to a meeting of experts may require tables, a report, graphs and diagrams, showing great detail. A group of shareholders may be interested in seeing only a graph which shows whether sales are rising or falling.

Frequently the first stage in presenting figures is to produce a table. Often this will be constructed before a report is written or graphs and diagrams drawn, because these will be based on the information contained in the table.

4.2   Tables

The purpose of tables is to facilitate the understanding of complex numerical data. Therefore a table should be as simple and unambiguous as possible.

There are a number of types of table, examples of which can be seen in government publications (such as *Financial Statistics*, the *Annual Abstract of Statistics*, *Economic Trends*, *Social Trends*), journals and magazines (such as the bank reviews and *The Economist*) and in company reports. Tables can be classified as:

1   informative tables providing a statistical record,
2   reference tables containing summarised information (backed up by more complex tables providing a complete analysis),
3   complex tables showing a number of columns and rows which are interrelated with sub-totals and totals divided into a number of categories.

Tables are used:

● to present the original numerical data in an orderly manner,
● to show a distinct pattern in the figures,
● to summarise the numerical data,
● to provide information which may help to solve problems.

THE CONSTRUCTION OF TABLES

The following points are important.

● All tables should have a title which gives a clear and concise indication of the contents.
● The original source of the data must be included (usually below the table).
● Column and row headings should be brief but self-explanatory.
● Units of measurement should be shown clearly.
● Approximations and omissions should be explained in table footnotes.
● A vertical arrangement of figures is generally preferable to a horizontal arrangement because columns of figures are easier to compare than rows.
● Double lines, or thick lines, can be used to break up a large table and make it easier to read.
● Two or three simple tables are often better than one very large table.
● Sets of data which are to be compared should be close together.
● Derived statistics, such as percentages and averages, should be beside the figures to which they relate.

Tables re-present data in order to emphasise certain features and to omit irrelevant detail, therefore it is important to be selective about the data that is put into a table and the number of tables produced.

When there are only a few variables involved tabulation can be carried out by hand quickly and easily. However, when there are a number of variables and combinations required with cross-tabulations, then the information can be fed into a computer. A computer will produce large numbers of tables very quickly if required and in a short time tables can be produced that can involve weeks of work to analyse and interpret. Therefore it is important to be clear about the aims and objectives of presenting the data before vast numbers of useless tables are produced.

4.3 Classification

Discrete and continuous variables

Before data can be tabulated, interpreted and presented it must be classified. **Classification is the process of relating the separate items within the mass of data collected to the definition of various categories.**

Every item of data has characteristics, some of which are measurable attributes or variables (such as weight) and some of which are non-measurable attributes (such as beauty). Measurable variables can be of two types: discrete variables and continuous variables.

Discrete variables are measured in single units (that is, they are countable things such as people, houses, cars). However, statements are made about discrete variables which appear to contradict this: for instance, 'the average family has 2.3 children' and 'on average people live in 1.2 rooms'. In fact it is difficult to give meaning to a fraction of a child or a room, but when figures are averaged these results occur.

Continuous variables are in units of measurements which can be broken down into definite gradations. Examples include temperature in decimals of a degree, height or length in decimals of a centimetre or fractions of an inch. However, in practice, continuous variables are often converted to a discrete form by expressing values to the nearest appropriate unit of measurement. It is difficult and unnecessary to distinguish small differences.

Class intervals

In producing a table of a frequency distribution (see Section 4.4), class intervals have to be shown clearly and unambiguously. There are many different methods of doing this.

Where a **discrete variable** is involved the following method can be used:

(i) Number of people
     100–199
     200–299
     300–399
     400–499

There cannot be any confusion between these class intervals. It is clear that the

199th person will be in the first class, and the 200th person will be in the second class. Other forms used include:

(ii) No. of people	(iii) No. of people
100–200	100–
200–300	200–
300–400	300–
400–500	400–

These class intervals are less clear. In (ii) it is not clear into which class 200 should fall, or 300 or 400. In each case there are two classes which could include these figures. In (iii) it has to be inferred where the classes end, particularly the last class.

With **continuous variables** the level of approximation or rounding needs to be shown clearly. For example, the height of people can be shown to very precise levels of measurement such as a millimetre or a fraction of an inch. In practice these heights are likely to be the nearest centimetre or to the nearest inch:

(i) Height (to the nearest centimetre)
  1.40 metres–1.59 metres
  1.60 metres–1.79 metres
  1.80 metres–1.99 metres

Within these class intervals, someone who is 1 metre 59 centimetres tall will be in the first class. Somebody else who is 1 metre 59.5 centimetres will be in the second class. The 'rules' of approximation and rounding are followed.

These class intervals can be shown in the form:

(ii) Height	(iii) Height
1.40 m–1.60 m	1.40 m but less than 1.60 m
1.60 m–1.80 m	1.60 m but less than 1.80 m
1.80 m–2.00 m	1.80 m but less than 2.00 m

These forms show continuous variables, but not the extent of the approximations used. 1.40 m but less than 1.60 m could mean up to 1 metre 59.99 centimetres or up to 1 metre 59.49 centimetres. However, these forms can be used if the level of approximation is given.

Open-ended class intervals

Open-ended classes normally occur at the beginning or end of a frequency distribution, and can cause problems in constructing class intervals.

OPEN-ENDED CLASS INTERVALS

Age of employees
(in years)
under 20
20 but under 40
40 but under 60
60 and over

In deciding the extent of the open-ended class interval (which is often necessary to carry out calculations), there are a number of points to be considered.

● The extent of adjacent class intervals may be relevant. In the example, because the third class interval is 40 but under 60, it could be argued that it would be consistent to make the last class interval 60 but under 80.
● There may be practical factors involved in classification. In the example, if the school-leaving age is 16 then the probability is that the first class will be 16 but under 20. Also in the example, if the firm involved strictly enforced retirement at 65, the upper class would be 60 but under 65.

Decisions on the extent of open-ended class intervals are a matter of judgement, so it is useful to state (in a footnote or in the report) the reasoning behind the decision which has been taken.

4.4   Frequency distributions

These show the frequency with which a particular variable occurs.

For example, a traffic survey may show the following flow of vehicles passing a particular point during an hour:

Vehicles	Frequency
Cars	45
Lorries	22
Motor cycles	6
Buses	3

In practice the observer may have had a list of these types of vehicles and put a line or mark against the appropriate category when a vehicle passed:

Vehicles		Frequency
Cars	₩₩ ₩₩ ₩₩ ₩₩ ₩₩ ₩₩ ₩₩ ₩₩ ₩₩	45
Lorries	₩₩ ₩₩ ₩₩ ₩₩ II	22
Motor cycles	₩₩ I	6
Buses	III	3

A survey of the number of people per household in a particular street of 20 houses may produce the following results:

Households	No. of people	Households	No. of people
A	4	K	3
B	3	L	2
C	2	M	3
D	6	N	2
E	3	O	2
F	1	P	3
G	4	Q	5
H	3	R	4
I	2	S	5
J	2	T	4

These results can be put into a table in the form of a frequency distribution:

No. of people per household	No. of households
1	2
2	5
3	6
4	4
5	2
6	1
	20

This is a **grouped frequency distribution** with the frequencies grouped according to the number of people per household.

In frequency distributions a decision has to be made about the grouping or classification used. This is not a problem in the above example, but it is more difficult with very large amounts of data.

If a table is constructed from the wage list of a firm employing 1000 people it may be necessary to produce a summary in the form of a frequency distribution (Table 4.1):

Table 4.1 Frequency distribution: wage list of a firm employing 1000 people.

Classes (weekly wages, to the nearest £)	Frequency (no. of employees earning wages within these classes)
240 to 259	15
260 to 279	40
280 to 299	140
300 to 319	620
320 to 339	150
340 to 359	35
	1000

This is a grouped frequency distribution with the variable values grouped into intervals to provide a summary which clarifies the distribution of wages within the firm. For instance, it can be seen from the table that 91% of the employees earn between £280 and £339 a week, with only a relatively few people earning less or more than these limits.

From a table like Table 4.1 it is possible to see a structure in the figures which might be more difficult to identify by looking at a list of a thousand wages.

4.5 Reports

Reports form an early stage of presentation of data, produced alongside or immediately after tables have been constructed.

Reports can be used:

- to explain the background to the collection of data – the way the information has been collected, whether it has been as a result of a primary survey or come from secondary data;
- to explain reasons for producing some tables and not others;
- to interpret the information contained in tables and other forms of presentation;
- to emphasise points of importance.

Textual reports are often the simplest method of presenting data and the easiest to understand for people not used to assimilating facts from tables. However, reports can include too many figures for them to be assimilated easily, so that a combination of tables with a report may provide the clearest form of presentation.

The main features of a good report are:

- accuracy,
- brevity,
- clarity.

4.6 Histograms

Once tables have been constructed and a report written, the next stage in presentation is to produce diagrammatic illustrations of the data.

A **line chart** is the simplest diagram used to represent a frequency distribution, where the length of each line is proportional to the frequency.

LINE CHART

In the traffic survey in Section 4.4 the observations showed:

Vehicles	Frequency
Cars	45
Lorries	22
Motor cycles	6
Buses	3

A line chart can be drawn to represent this data. The length (or height) of each line represents the frequency (Figure 4.2):

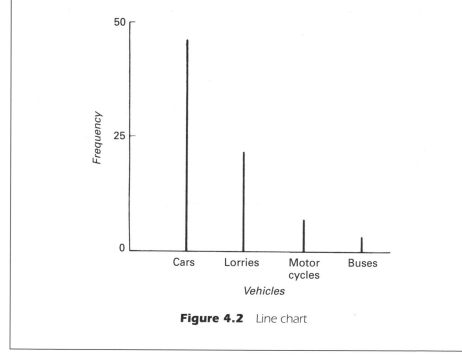

Figure 4.2 Line chart

The **histogram** is an extension of the line chart. The word 'histogram' is derived from the Greek *histos*, or mast, so it is a mast diagram or bar diagram. **A histogram consists of a series of blocks or bars each with a width proportional to the class interval concerned and an area proportional to the frequency.**

Whereas in a line chart the length of a line is proportional to the frequency, it is the *areas* of the blocks which must be proportional in a histogram.

This can be seen by amalgamating the last three groups in the frequency distribution in Table 4.2. This will make a class interval of 250 units to 279 units with a frequency of 33 operatives. This is shown by drawing a block which is three times the width of the others and a third the height (Figure 4.4).

HISTOGRAM

In a survey of the output of machine operatives in a factory, the results shown in Table 4.2 were obtained:

Table 4.2   Output of machine operatives.

Output (units per operative)	Number of operatives
200–209	5
210–219	14
220–229	17
230–239	29
240–249	42
250–259	21
260–269	10
270–279	2

In a histogram this frequency distribution would appear as shown in Figure 4.3.

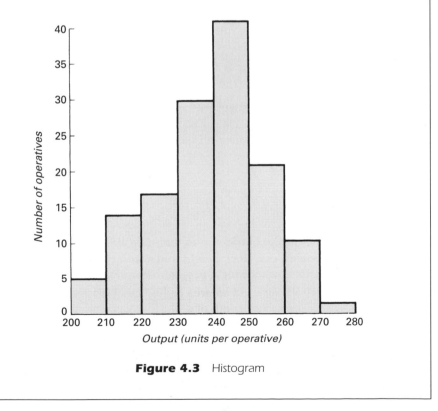

Figure 4.3   Histogram

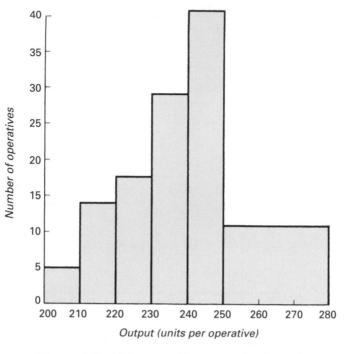

Figure 4.4 Histogram with uneven class intervals

The area of a histogram bar is found by multiplying the width of the bar (the class interval) by the height (the frequency). Therefore,

the area of the last block in Figure 4.4 is:
$30 \times 11 = 330$
the areas of the last three blocks in Figure 4.3 are:
$10 \times 21 = 210$
$10 \times 10 = 100$
$10 \times 2 = 20$
330

Therefore, **if the areas of the blocks are to maintain their proportional relationship with the frequencies, the block height and width must be changed proportionately.** It follows that, whenever a grouped frequency distribution includes uneven class intervals, the areas of the blocks must be kept in proportion.

THE CONSTRUCTION OF HISTOGRAMS

The following points are important.

- The horizontal axis is a continuous scale including all the units of the grouped class intervals.

- For each class in the distribution a block (or vertical rectangle) is drawn with width extending from the lower class limit to the upper limit.
- The area of this block will be proportional to the frequency of the class.
- If the class intervals are even throughout a frequency table, then the height of each block is proportional to the frequency.
- There are never gaps between histogram blocks because the class limits are the true limits in the case of continuous data and the mathematical limits in the case of discrete data.

4.7 Frequency polygons

A frequency polygon is drawn by joining up the mid-points of the tops of histogram blocks. It is usually drawn with straight lines, and the resulting diagram, taken with the horizontal axis, is a 'many-sided figure' or polygon.

Figure 4.5 shows a frequency polygon drawn from the histogram based on the data in Table 4.3. In the case of the first and last class intervals, the line is extended beyond the original range of the variable (see Figure 4.5). This is because the area under the polygon should be the same as that in the histogram. Only if each triangle cut off the histogram is compensated for will this requirement be met. This can be seen in Figure 4.5 by the lettered triangles.

In theory a frequency polygon is always drawn by constructing the histogram first; in practice, each point can be located by reference to the frequency and the mid-point of the class interval. Therefore, using the data from Table 4.3, the frequency polygon in Figure 4.6 can be constructed.

Table 4.3 Overtime pay.

Overtime pay (to the nearest £)	Number of employees
0–4	4
5–9	10
10–14	5
15–19	4
20–24	2
	25

Histograms and frequency polygons enable the properties of distributions to be examined, and various forms of distribution can be compared in a general way (see Chapter 7 for comparisons). However, neither the histogram nor the frequency polygon gives a very accurate picture of a frequency distribution. The histogram implies that the frequencies are the same throughout the class interval, when they may not be; the frequency polygon implies that sharp, angular differences occur between the mid-points of the class intervals when this may not be true.

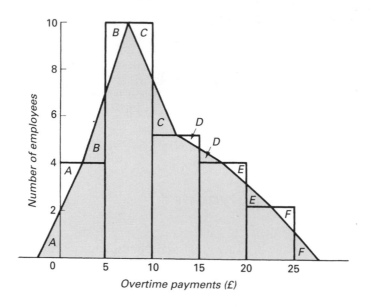

Figure 4.5 Frequency polygon constructed from histogram

It is likely that a more accurate idea of the distribution would be obtained if there were smaller class intervals and more observations. In other words the closer the class intervals are to the original set of figures the greater the accuracy obtained. This may be at a cost of a loss of summarisation and clarity.

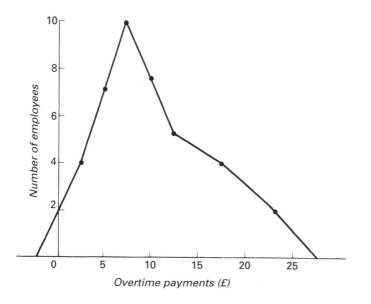

Figure 4.6 Frequency polygon constructed direct from data

4.8 Frequency curves

A frequency curve is formed by:

- smoothing out a histogram or frequency polygon, or
- using smaller class intervals and more observations to smooth out the line of the curve.

As the number of class intervals and frequencies are increased so the polygon and histogram move more closely towards a curve (Figure 4.7):

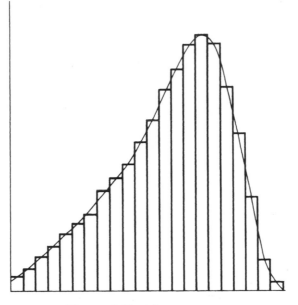

Figure 4.7 A frequency curve

The importance of the frequency curve, like the histogram and polygon, is that it can provide a clear picture of the 'shape' of a distribution. With a small distribution the shape of a distribution may be seen easily by looking at the figures, but with large amounts of data it may be more difficult. The frequency curve can provide a summary at a glance and make an immediate impact.

More than one frequency curve can be plotted on the same axis for comparison (see Figure 4.8).

From the two curves it can be seen that:

- in general, company A made lower overtime payments than company B;
- the highest level of overtime is paid by company A, as well as the lowest;
- company B has a greater number of employees earning overtime than company A (the area under curve B appears to be larger than the area under curve A);
- most of company B's overtime payment is made at the top end of the

overtime scales, while company A's overtime payments are at the lower end of the scale;

- there is some overlap of overtime payment.

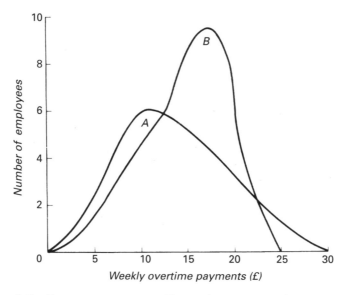

Figure 4.8 Frequency curves: weekly overtime payments by two companies

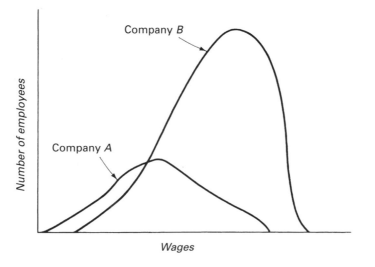

Figure 4.9 Frequency curves for Assignment 4.3

ASSIGNMENTS

1   Draw a histogram from the data in Table 4.1. Briefly comment on the distribution.
2   Write a short report on the problems of classification.
3   Comment on the differences in the two distributions shown in Figure 4.9.
4   Discuss, with examples, the aims of statistical presentation.
5   (i)   Draw up a table to present the information that is likely to be collected from Assignments 3.2 and/or 3.3.
    (ii)   Draw up a table to show the main tasks carried out during a person's working day and the time spent on each task.

⬡5 How to present statistics (2)

OBJECTIVES

- To provide an understanding of the construction and use of bar charts, pie charts, pictograms, cartograms and strata charts, Gantt charts, break-even charts, Z-charts and Lorenz curves
- To provide an understanding of the construction and use of graphs
- To analyse the problems of presentation and perception

Tables, reports and frequency distributions are the basic methods of presenting raw data. The next stage is to present data in a pictorial form which will make an immediate impact, illustrate the information and bring out the salient points. Pictorial presentation falls into two main categories:

1 charts,
2 graphs.

5.1 Bar charts

Bar charts are among the most popular forms of pictorial presentation. There is a range of commonly used bar charts.

The simple bar chart

The simple bar chart should not be confused with the histogram. It is more like a line chart than a histogram (see Section 4.6) because **it is the height (or length) of each bar that represents the data; the width and area of the bar are not important** because they are not drawn in proportion to any data (as they are in the histogram). For this reason all the bars on any particular bar chart are the same width.

Simple bar charts (like Figure 5.1) can be used to illustrate only simple

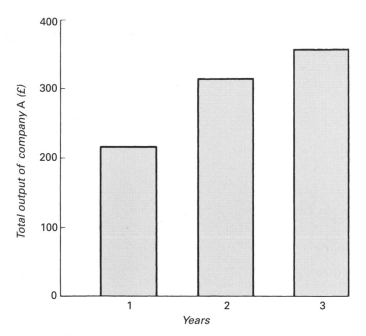

Figure 5.1 The simple bar chart: output of a company over three years

pieces of information but can illustrate information clearly and provide an immediate visual impact. Figure 5.1 shows that the total output of company *A* has increased over the three years but has increased more slowly between year 2 and year 3 than between year 1 and year 2. The bar chart makes comparison between years easy.

The bar chart can be drawn with horizontal bars (Figure 5.2), if this improves the visual impact of the chart.

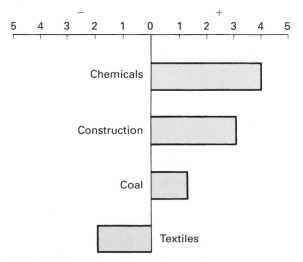

Figure 5.2 Horizontal bar chart: changes in output over two years

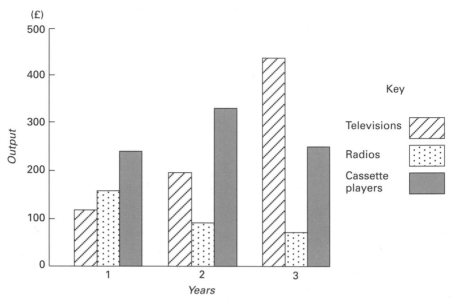

Figure 5.3 The compound or multiple bar chart: output of an electrical goods company

The compound or multiple bar chart

A multiple bar chart (see Figure 5.3) is useful for comparing a number of items within say a year, as well as comparing the items between the years.

Component bar charts

Component bar charts are useful to show the division of the whole of an item into its constituent parts. Figure 5.4 shows the output of an electrical goods company over three years. The chart shows that total output in each year has risen. The output of television sets has increased but the output of radio sets has declined while the output of cassette players has fluctuated.

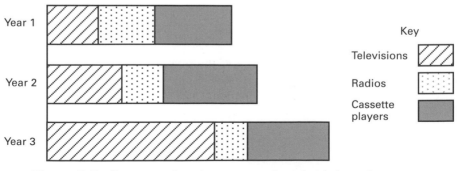

Figure 5.4 Component bar chart: output of an electrical goods company

This type of chart enables a comparison of the total output to be made easily, while the compound bar chart emphasises the comparisons between items.

Percentage component bar chart

In this type of chart (Figure 5.5) the bars represent 100% and therefore remain the same length. The components change to represent the *percentage* they make up of the total.

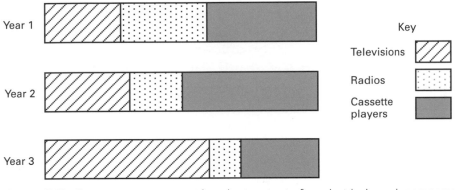

Figure 5.5 Percentage component bar chart: output of an electrical goods company

Figure 5.5 shows the changes in the way the total output is made up (its components), but does not show the fact that total output has risen over the three years. The chart shows that televisions have become a larger proportion of total output, while radios and cassette players have become a smaller proportion.

5.2 Pie charts

A pie chart is a circle divided into sectors to represent each item or variable (pie or pi from the Greek π, πr^2 the area of a circle, or pie as in apple). Each sector of the circle should have an area proportional to the proportion that the variable makes of the whole. For example, if the quantity of a variable is 15 units out of a total of 45, then the variable must occupy $\frac{15}{45}$ of the area of the circle. Since a pie chart can be constructed with any radius, the easiest way to divide it into sectors is using angles: that is, taking proportions of a complete turn (360°). Therefore this variable will occupy:

$$\frac{15}{45} \times 360 = 120°$$

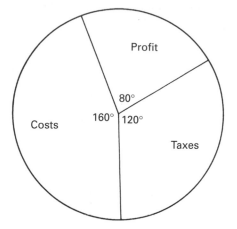

Figure 5.6   Pie chart: company finances

So, if the finances of a company are broken down as:

	£ million
Costs	20
Profit	10
Taxes	15

a pie chart drawn to represent this information should be divided in proportion:

$$\frac{20}{45} \times 360 = 160°, \quad \frac{10}{45} \times 360 = 80°, \quad \frac{15}{45} \times 360 = 120°$$

The three sectors of the circle will be 160°, 80°, and 120° (Figure 5.6).

Often pie charts are presented with the areas shaded or coloured and a key provided (Figure 5.7):

Pie charts are useful where there are a few items which make up propor-tions of a whole and where the proportions are more important than the

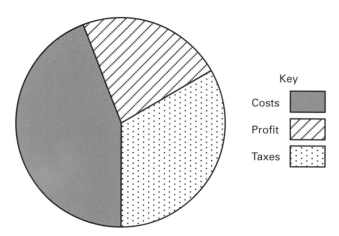

Figure 5.7   Shaded pie chart: company finances

numerical values. For instance, shareholders may be more interested in the proportion of profit than the actual values; they want to know the size of their share of the cake.

5.3 Comparative pie charts

Comparative pie charts are used to compare two sets of data, usually two sets of the same items over time. The areas of the circles must be in proportion to the totals of the data.

For example, if the finances of a company over two years are:

	Year 1 (£ million)	Year 2 (£ million)
Costs	20	20
Profit	2	10
Taxes	8	15
	30	45

A pie chart is constructed for year 1 with a radius of say 2 cm. The area of a circle is πr^2. If $r = 2$ cm $r^2 = 4$ cm^2 so £30 million is represented by a circle of area $\pi \times 4$ square centimetres (cm^2). £10 million would be represented by a circle of area $\pi \times 4 \div 3 = \pi \times 1.33$ cm^2, £45 million would be represented by a circle of area $(\pi \times 4/30) \times 45 = \pi \times 6$ cm^2.

If the radius of the pie chart representing year 2 is r, the area in square centimetres will be πr^2. The area has to be $\pi \times 6$ cm^2. Therefore

$$\pi r^2 = \pi \times 6$$
$$r^2 = 6$$
$$r = 2.4 \text{ cm}$$

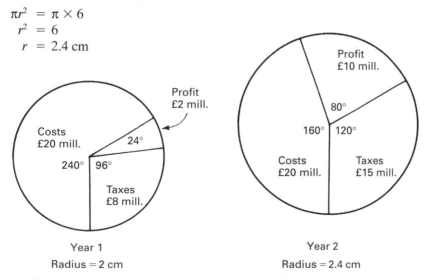

Year 1
Radius = 2 cm

Year 2
Radius = 2.4 cm

Figure 5.8 Comparative pie charts: company finances over two years

There is no need to substitute the numerical value for π.

Therefore the first pie chart will have a radius of 2 cm and the second pie chart a radius of 2.4 cm with the appropriate sectors for the three variables (Figure 5.8).

While the sectors can be compared relatively easily, the areas of the two circles are too similar to each other to make comparison easy.

Pie charts are useful when:
- there are few variables to be included;
- these variables make up proportions of a whole figure;
- the numerical values are less important than the relative proportions of the whole;
- the aim is to provide a strong visual impact.

Disadvantages of pie charts
- They can involve long calculations.
- They do not provide information on absolute values unless figures are inserted in each sector.
- The sectors cannot be scaled against a single axis as the values can on a bar chart.
- To compare two totals, the areas of the pie charts should be in proportion; the fact that the area of one circle is larger than the area of the other circle may not be very clear visually.

5.4   Pictograms

Pictograms are *pict*orial dia*grams*. They are pictures to represent data. The objective in using them is to provide an immediate visual impact and therefore they should be kept simple. Pictograms cannot give detailed information, but can show trends, comparisons and totals.

Figure 5.9 shows that the number of cars produced by this company is falling, and at the same time the number of workers is falling: at the same speed between year 1 and year 2, but faster between year 2 and year 3. In year 1 the ratio of workers to cars produced was $3:1$, in year 2 it was $3:1$ and in year 3 it was $2:1$.

Advantages of pictograms
- They can be an attractive method of representing data.
- They make it easy to appreciate fluctuations in variables.
- They make an immediate visual impact.

Disadvantages of pictograms
- It may not be easy to estimate the precise numbers involved.
- They may be misleading unless the symbols used are always the same size. It is easier to compare variables by increasing the number of symbols rather than the size or area of the symbol.
- They may be used for showing only relatively simple information.

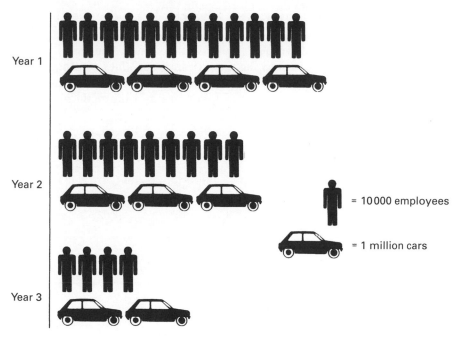

Figure 5.9 Pictogram: number employed compared with the output of cars

5.5 Comparative pictograms

Comparative pictograms can be used to compare relationships by using symbols and pictures of different sizes. As in the pie chart, the area of the symbol should be in proportion to the quantity of the items being compared. Figure 5.10 shows a comparative pictogram to represent £10 000 and £40 000. Square B is four times the area of square A, to represent an amount of money four times larger.

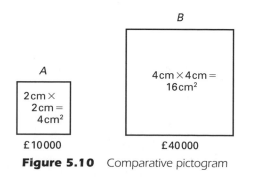

Figure 5.10 Comparative pictogram

5.6 Cartograms or map charts

These are maps onto which graphs, symbols, pictograms, flags and so on are superimposed to represent various factors. For instance, flags or other symbols can be placed on maps to show outlets for a company's goods. Also, cartograms are used to show the main industries of a country and the main imports and exports.

Any of the other forms of presentation discussed in this chapter (such as pie charts or bar charts) can be superimposed on to a map where this is appropriate. Cartograms can provide an attractive and easily understood visual form of representation where geographical location is important (see Figure 5.11).

Figure 5.11   Cartogram: company sales of three products (by proportion) in regions of England and Wales

5.7 Strata charts

Strata charts, also called layer charts or band curve graphs, are given this name because each band or strata is placed successsively on top of the previous one. Totals are cumulative, so that each element of the whole is plotted one above another. These charts can provide immediate evidence of the relative importance of the various constituents of the total.

In Figure 5.12 the top line shows the total production costs. It is relatively easy to see the amount made up by each layer. Overheads are a larger propor-

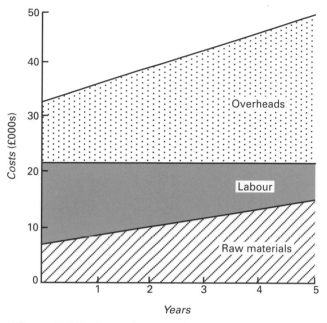

Figure 5.12 Strata chart: production costs of a company

tion of total costs than labour or raw materials by year 5. It is more difficult to compare layers or strata which are close to each other in height, or to read off the amounts for each item.

5.8 Graphs

In theory, a graph consists of a grid with four quadrants, with zero (the origin) at the centre (Figure 5.13). Curves (or lines) are drawn on the grid to illustrate the relationship between two variables. Usually, however, the negative axes are not drawn on a graph because, although variables can be negative as well as positive, they are usually positive. A graph is a form of pictorial presentation and the word can be associated with the word 'graphic' or a 'vivid representation'.

The vertical axis, or y axis (the positive vertical axis), is scaled in units of the dependent variable. The horizontal axis, or x axis (the positive horizontal axis), is scaled in units of the independent variable. **The independent variable is the variable which is not affected by changes in the other variable. The dependent variable is the variable which is affected by changes in the other variable.** For example, changes in advertising expenditure may affect sales, but the level of sales will not directly affect advertising expenditure. Therefore advertising is the independent variable and sales the dependent variable (Figure 5.14).

In practice it is not always easy to decide which variable is dependent and which independent. For instance, changes in the costs of a firm may affect out-

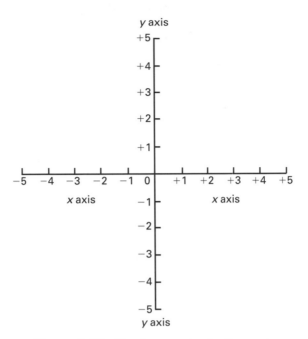

Figure 5.13 The four quadrants of a graph

put and output may affect costs. From different viewpoints they may be dependent on each other.

Normally the independent variable is chosen before the information is collected and therefore it is stated in units or class intervals or periods of time. The dependent variable is the number or frequency in each class interval or period of time.

The curve of a graph is usually drawn as a freehand curve, but may be

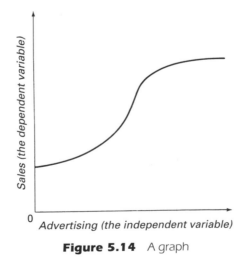

Figure 5.14 A graph

drawn as a series of straight lines. Strictly, if the dependent variable is discrete then the points should be joined by straight lines. If the dependent variable is continuous then the points should be joined by a smooth curve.

In fact a smooth curve is often used with discrete as well as continuous variables. However, it is only with continuous variables that values between points can be read off (interpolated), because with a discrete variable there is no information between one point and the next.

Also, with continuous variables it is possible to extrapolate to find a value outside the given range, by extending the graph on the assumed trend (see Chapter 11).

The position of any point on the curve of a graph is decided by reference to the x and y axes. **The points where the variables intersect are called 'bearings' or 'co-ordinates'.** A set of points is built up and these are joined to form the curve. Like reading a map, on a graph a point is fixed by reading out from the origin (where the x and y axis intersect at 0) along the horizontal axis and drawing a vertical line up from this point and then reading up from the origin on the vertical axis and drawing a horizontal line along from this point (as in map reading this can be described as: 'along the corridor and up the stairs').

The data in Table 5.1 can be used to plot six points from which a graph can be drawn (Figure 5.15). The graph can then be used to interpolate values between these points.

Table 5.1 Monthly sales.

Months	Sales of company A (£000s)
January	100
February	250
March	400
April	480
May	420
June	340

From Figure 5.15 it is possible to interpolate that the level of sales in the middle of March was about £480 000. If it is assumed that the downward direction of the curve continues then it is possible to extrapolate that sales will be down to a level of £155 000 by July. However, *this is a dangerous assumption* because the trend could have changed due to a range of factors and the period of the graph does not provide sufficient evidence of the trend (see Chapter 11).

On a graph the vertical axis (or scale) should always start at zero, otherwise a false impression may be created. If it is not practical to have the whole scale running from zero then the scale can just cover the relevant figures provided that zero is shown at the bottom of the scale and a definite break in the scale is shown. This break can be shown in two ways:

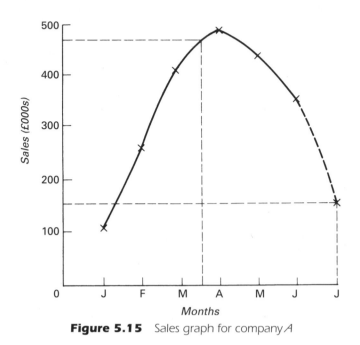

Figure 5.15 Sales graph for company *A*

- the vertical axis can be broken or zigzagged to show that some of the empty space has been compressed (Figure 5.16);
- the vertical scale can be broken by two jagged lines running across the diagram to indicate that a portion of the space has been omitted (Figure 5.17).

When the method of compiling or calculating the figure under review has been changed, this should be shown by a break in the axis *and* in the graph (Figure 5.18). This happens with index numbers when a new series is developed with a new base date and weighting system (see Chapter 9).

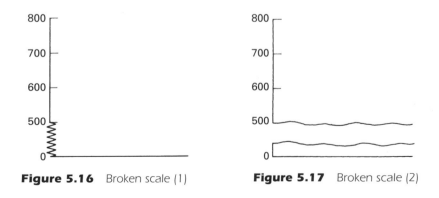

Figure 5.16 Broken scale (1)   **Figure 5.17** Broken scale (2)

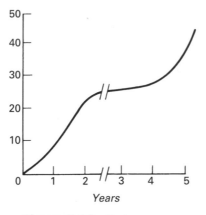

Figure 5.18 Broken curve

5.9 Semi-logarithmic graphs

The semi-logarithmic graph (or semi-log or ratio-scale graph) is used to show the rate of change in data, rather than changes in actual amounts (which are shown on natural scale graphs).

Usually only one axis (the *y* or vertical axis) is measured in a logarithmic or ratio scale. This is why the graph is called a *semi*-log graph.

The important factor on a semi-log graph is the degree of slope of the curve. The curve of the usual graph measures the magnitude at any point, while the log graph shows at any point the percentage change from the last point (see Table 5.2, Figure 5.19 and Figure 5.20).

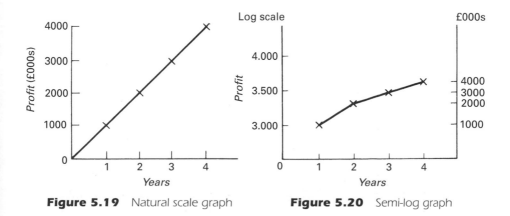

Figure 5.19 Natural scale graph **Figure 5.20** Semi-log graph

Table 5.2 Semi-log table.

Years	Profit (£000s)	Increase in profit over previous year (%)	Log numbers of the profit figures
1	1000	100	3.0000
2	2000	100	3.3010
3	3000	50	3.4771
4	4000	$33\frac{1}{3}$	3.6021

Notice that in Figure 5.20 it is the log numbers that are charted on the vertical axis. This graph shows that, although profits are rising, they are rising at a declining rate, while the natural scale graph (Figure 5.19) suggests a constant expansion of profits. They are both 'correct'; they simply show different aspects of the same information.

There are two methods of producing a semi-log or ratio graph:

- by plotting the graph on special graph or ratio-scale paper,
- by plotting the log of the data on the vertical axis.

Interpreting semi-log graphs
- The curve with the greatest slope is the one with the greatest rate of increase.
- The slope of the curve indicates the rate at which the figures are increasing.
- If the 'curve' is a straight line, the rate of increase remains constant.
- If the absolute increase is constant the curve will become progressively less steep (as in Figure 5.20).

5.10   Straight-line graphs

Straight-line graphs occur when the arithmetical relationship between two sets of data is 'direct variation', in which a change in one variable is matched by a similar change in the other variable. For example, at a given time, one British pound is worth a specific number of US dollars and therefore as pounds are converted to dollars the relationship remains in direct variation. If £1 = $2, then £10 = $20 and £100 = $200.

Hourly paid work involves a direct variation between pay and hours. If the rate of pay is £9 an hour, then:

20 hours' work earns 20 × £9 = £180
40 hours' work earns 40 × £9 = £360
60 hours' work earns 60 × £9 = £540

This can be shown on a graph (see Figure 5.21).

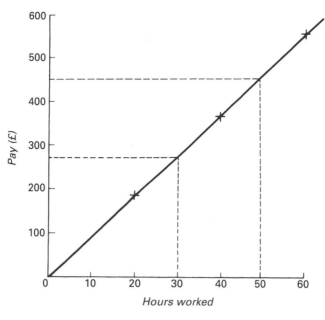

Figure 5.21 Straight-line graph: direct variation

The advantage of drawing the graph is that it is possible to interpolate information from it. In fact the graph can be used as a ready reckoner:

30 hours of work earns £270,
50 hours of work will earn £450.

The straight-line graph is an application of graphical methods. Other pictorial methods are also applied to particular circumstances and uses. The following examples are some of these.

5.11 Gantt charts

The Gantt chart is often used in production as a progress chart. Usually it consists of two horizontal bar charts for each period of time. One bar may indicate the planned production or running time and the other bar the actual figures. Any discrepancy between the two reveals a loss of production.

In Figure 5.22 the horizontal scale shows 100% for each day, which indicates the total possible production. The thinner line or bar shows the planned production based on the variable factors in production. The thicker bar shows the actual production achieved each day during a particular week.

The chart shows that the amount of production planned is much the same on Monday, Tuesday and Wednesday and rather less on Thursday and Friday. Actual production has equalled planned production only on Monday and fell well short on Friday.

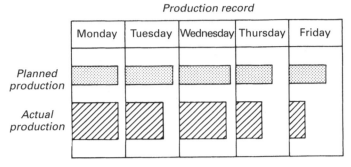

Production record

	Monday	Tuesday	Wednesday	Thursday	Friday

Planned production

Actual production

Figure 5.22 A Gantt chart

This type of information may be used as the basis for action, for instance to increase actual output, or to reduce target production to realistic levels. These charts can be used to control the use of machines and in various other aspects of production.

5.12 Break-even charts

Break-even charts show the profit or loss for any given output. The simplest chart shows two curves or straight lines, one showing the relationship between revenue and output, the other the relationship between cost and output.

For example, a particular make of radio has variable costs of £30 per unit, overall fixed costs of £100 000 and a selling price of £60 a unit. This produces a table of output, costs and revenue (Table 5.3). The data in Table 5.3 can be used to construct a break-even chart (Figure 5.23). Where the two lines cross is the break-even point (at about 3500 units of output). At this output, revenue covers

Table 5.3 Break-even table.

Units (thousands)	Costs (£)	Revenue (£)
1	130 000 (30 + 100 000)	60 000
2	160 000	120 000
3	190 000	180 000
4	220 000	240 000
5	250 000	300 000
6	280 000	360 000

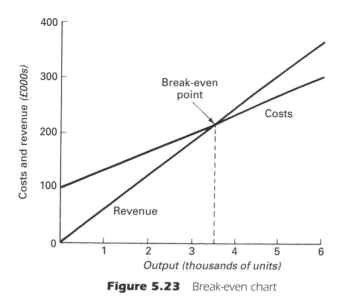

Figure 5.23 Break-even chart

costs, while below this output there is a loss because costs are greater than revenue and above this point there is a profit, because revenue is greater than costs.

5.13 The Z chart

The Z chart consists of three graphs plotted together on the same axis. The three graphs are:

- the original data,
- the cumulative total,
- the moving annual total.

When plotted on a graph the three curves form the shape of a 'Z'. The curve of the original data shows the current fluctuations, the cumulative curve shows the position to date and the trend is indicated by the moving annual total. **This chart can be used to compare the basic data with trends of data,** such as sales. To calculate the moving annual total, figures are needed for two years (see the example on page 86).

Table 5.4 shows monthly sales for two years, cumulative monthly total for year 2, and moving annual total for year 2. Figure 5.24 shows the associated Z chart. The moving annual total is calculated by taking the total sales for year 1 and then adding the sales for each month of year 2 (in turn), while subtracting the sales of the corresponding month of year 1.

Table 5.4 Monthly sales.

Month	Monthly sales (£) (Year 1)	Monthly sales (£) (Year 2)	Cumulative monthly total (£) (Year 2)	Moving annual total (£) (Year 2)
January	25	30	30	735
February	45	50	80	740
March	60	50	130	740
April	65	70	200	745
May	80	80	280	745
June	85	90	370	750
July	90	100	470	760
August	100	110	580	770
September	75	100	680	795
October	60	70	750	805
November	35	50	800	820
December	20	35	835	835
Total	730	835		

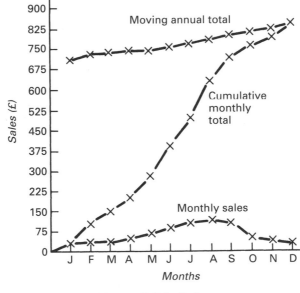

Figure 5.24 Z chart

5.14 The Lorenz curve

The Lorenz curve is a graphical method of showing the deviation from the average of a group of data. It is a cumulative percentage curve.

Table 5.5 Accumulated wealth (1).

Income (£000s) (1)	No. of people (f) (2)	Accumulated wealth (£000s) (3)	Cumulative wealth (£000s) (4)	(%) (5)	Cumulative frequency (f) (6)	(%) (7)
Less than 5	144	32	32	16	144	48
5–9.9	54	22	54	27	198	66
10–14.9	36	24	78	39	234	78
15–19.9	24	20	98	49	258	86
20–24.9	18	24	122	61	276	92
25–29.9	15	26	148	74	291	97
30–34.9	9	52	200	100	300	100

The curve is often used to show the level of inequality. For instance, it can show the number of people saving against the amount saved. The more equal the distribution of saving, the flatter the curve. If there was equality between the two variables, the curve would be a straight line, equal to the 'line of equal distribution'.

The Lorenz curve gives an immediate impression and it is used for comparison rather than as a quantitative measure of inequality.

Tables 5.5 and 5.6 and Figures 5.25 and 5.26 show comparison between the number of people and accumulated wealth. They show a comparatively high

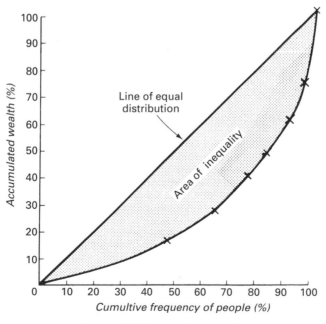

Figure 5.25 Lorenz curve (1)

Table 5.6 Accumulated wealth (2).

Income (£000s) (1)	No. of people (f) (2)	Accumulated wealth (£000s) (3)	Cumulative wealth (£000s) (4)	(%) (5)	Cumulative frequency (f) (6)	(%) (7)
Less than 5	50	25	25	12.5	50	16.66
5–9.99	46	20	45	22.5	96	32
10–14.9	45	26	71	35.5	141	47
15–19.9	39	31	102	51	180	60
20–24.9	34	22	124	62	214	71.33
25–29.9	42	31	155	77.5	256	85.33
30–34.9	44	45	200	100	300	100

(Figure 5.25) and low (Figure 5.26) degree of inequality. The curves are drawn by plotting the percentage of accumulated wealth (column 5) against the percentage of the cumulative frequency of people (column 7).

Figure 5.25 indicates that the distribution of wealth is not very equal. If there was 'equality' then 10% of the income earners would have 10% of the total wealth and 50% of the earners would have 50% of the total wealth and so on. Table 5.6 and Figure 5.26 show a high degree of equality between wealth and income earners.

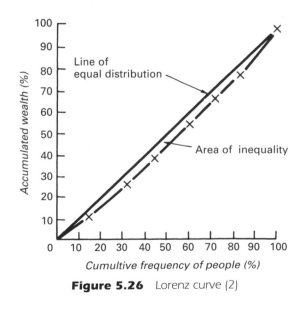

Figure 5.26 Lorenz curve (2)

5.15 Presentation and perception

Charts and graphs can illustrate information, emphasise the central points as well as areas of difference and similarity, and clarify the meaning of complicated data.

Presentation involves a process of summarisation and simplification. One problem with this is that it is possible to over-simplify and to present information in an ambiguous or misleading way.

The problems of presentation can include:

- difficulties of perception,
- problems of distortion and deception.

Perception

It is not always possible to be sure what is seen.

PERCEPTION

Example 1
In Figure 5.27, which segment of the line *ABC* is the greater: *AB* or *BC*?

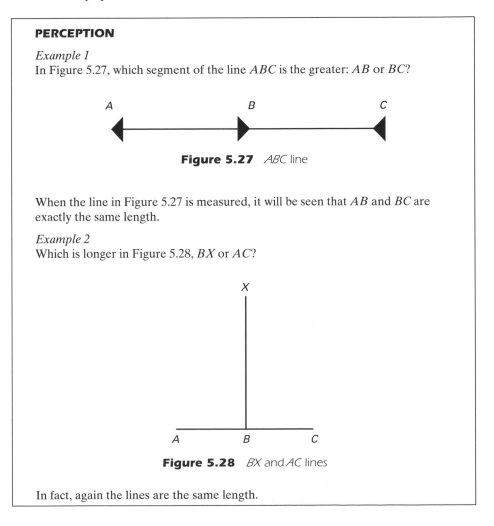

Figure 5.27   *ABC* line

When the line in Figure 5.27 is measured, it will be seen that *AB* and *BC* are exactly the same length.

Example 2
Which is longer in Figure 5.28, *BX* or *AC*?

Figure 5.28   *BX* and *AC* lines

In fact, again the lines are the same length.

Example 3

Then there is the famous young woman, old woman ambiguity: is Figure 5.29 a picture of a young woman or an old woman?

Figure 5.29   Young woman, old woman

In experiments 60% of people have seen the younger woman, 40% the older woman. In fact both women are in the picture.

These three examples highlight the point that it is not possible to take the perception of illustrations and diagrams for granted.

Distortion and deception

Any illustrations may deceive by misrepresenting information.

Examples include bar charts with bars of different width, pictograms with different size symbols which are not in proportion, and graphs which do not keep to the basic rules.

DISTORTION IN GRAPHS

Not starting the *y* axis at zero

The distortion in Figure 5.31 makes the curve appear steeper than it does in Figure 5.30 where the origin is shown; this gives the impression that sales have increased more rapidly than they have in fact.

Figure 5.30   Actual sales

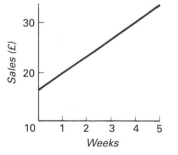

Figure 5.31   Sales not showing the zero

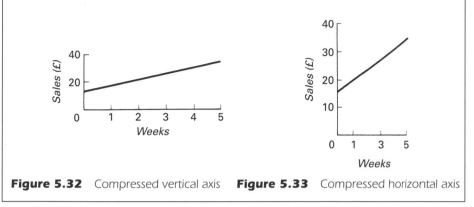

Compressing the vertical or horizontal axis

Compressing an axis will flatten or sharpen the angle of the curve (see Figures 5.32 and 5.33).

Figure 5.32 Compressed vertical axis   **Figure 5.33** Compressed horizontal axis

Graphs may be distorted by not starting the *y* axis at zero, or by compressing the vertical or horizontal axis.

All forms of distortion and deception, or illustrations which encourage ambiguity of perception, work against the basic aims of statistical presentation, which are clarity of communication and accuracy in the presentation of information.

ASSIGNMENTS

1   Look in any government publication and select three different types of statistical presentation. Discuss the reasons why each particular type of presentation was chosen and whether other forms of presentation might have been just as good, or better.
2   Draw suitable diagrams and graphs to illustrate Table 1.1. Briefly justify your choice of diagram.
3   Discuss the problems of perception and distortion in the presentation of information. Comment on particular advertisements which make use of these 'concepts'.
4   Carefully and fully interpret the information shown in Figures 5.1, 5.2, 5.3, 5.4, 5.5, 5.6, 5.8, 5.9.
5   Discuss the advantages of the various forms of graph compared with histograms and bar charts.
6   Look at the annual report of a company and discuss critically the forms of statistical presentation used.

⬡₆ **Summarising data: averages**

6.1 The role of the average

Averages are measures of central tendency and measures of location. As a measure of central tendency an average provides a value around which a set of data is located. As a measure of location an average provides an indication of whereabouts the data is situated. An average price can give an indication of whether a particular commodity is likely to cost £1 or £10 or £100.

An average summarises a group of figures and represents it in the sense that the average provides an immediate idea about the group. An average can provide a description of a group of items so as to distinguish it from another group with similar characteristics.

Therefore an average is a way of describing data very concisely, and because of this there are a number of averages which can be used depending on the type of description required. **The three most commonly used averages are the arithmetic mean, the median and the mode.**

Averages are used all the time in everyday life and work. The statement that 'incomes have risen by 10% in the last year' is usually qualified by the word 'average': 'average incomes have risen by 10% in the last year'. The *average* consumer is said to buy more each year, the *average* rainfall has risen or fallen, the *average* height of people . . ., the *average* price of electrical goods . . . and so on.

Averages can be used in a number of ways.

An average can **summarise a group of figures, smoothing out abnormalities in a way that is useful for comparison.**

Example 1
In cricket, if a batsman has scored 1000 runs in 40 innings in all of which he is out, the scores in each innings may have varied between 0 and 200. The average would be 25 runs per innings (1000 divided by 40). This figure can be compared with the average achieved by other batsmen.

Example 2
Two firms may pay very widely varying wages to similar types of employees. In one firm, wages may vary between £245 a week and £425; in the other firm, wages may vary between £285 and £350. Both firms may have an average wage for this type of employee of £300.

An average can provide a mental picture of the distribution it represents.

It may be recorded that a shop that is for sale has average weekly sales of £50 000. This provides an immediate idea of the size of the business, although sales might have been £100 000 in Christmas week and £5000 in the worst week.

An average can provide valuable knowledge about the whole distribution.

If the average wage in a factory is £300 and there are 1000 employees, then it can be deduced that the weekly wage bill is £300 000.

The word 'average' is used in daily conversation and is used loosely.

The statement 'I think that on average I use about ten gallons of petrol a week', is using the average as an estimate.

Averages can conceal important facts.

Two companies may both have average annual profits of £50 000 over the last five years. However, when their actual records are inspected it may be found that the companies' performances are very different:

	Year 1 £	Year 2 £	Year 3 £	Year 4 £	Year 5 £
Company A	110 000	80 000	35 000	20 000	5 000
Company B	5 000	20 000	35 000	80 000	110 000

In the same way, it is important to know not only the average but also other figures such as the minimum and maximum. An engineer designing a reservoir must know not only the average rainfall of the region, but also the maximum.

Therefore averages can provide the first stage of an investigation, but do not provide all the information required for many purposes.

6.2 The arithmetic mean

The arithmetic mean can be defined as the sum of the items divided by the number of them.

This is the average to which most people refer when they use the word 'average'.

Example 1
If five people have £15, £17, £18, £20, £30, respectively, the arithmetic mean is £100/5 = £20. That is, if £100 was shared equally between five people, they would have £20 each.

$$\text{The arithmetic mean} = \frac{\text{total value of items}}{\text{total number of items}}$$

$$= \frac{\Sigma x}{n}$$

where Σ is the sum of,
  x is the value of the items,
  n is the number of items.

The usual symbol for the arithmetic mean is $\bar{x}$ (x-bar or bar-x). Strictly, $\bar{x}$ is the symbol for the arithmetic mean of a sample and μ (mu or mew) is the symbol for the arithmetic mean of the population from which samples are selected. However, $\bar{x}$ is used very widely.

Another method of calculating the mean is to assume an average by inspection, find the deviation of the items from this assumed average, sum the deviations, average them and add or subtract this from the assumed average.

Notice that the formula used in Example 2 is:

$$\bar{x} = x \pm \frac{\Sigma d_x}{n}$$

where x represents the assumed mean and d_x represents the deviations from the assumed mean. In this case the assumed mean is £18. The deviations of the items from this assumed mean total +10. This is a total deviation of +10 over five items, averaging 2 per item. The 2 is added (in this case) to the assumed mean, to arrive at the same answer (£20) as the first method.

With large numbers of figures this method may be faster than the first method. Also it is useful to have an introduction to it, for further calculations.

Any figure chosen as the assumed mean would give the correct answer.

Example 2

If five people have £15, £17, £18, £20, £30, the average can be assumed to be £18. Deviations from this assumed mean will be:

Item (£)	Deviations
15	−3
17	−1
18	0
20	+2
30	+12
	−4 + 14 = +10

$$\bar{x} = 18 + \tfrac{10}{5}$$
$$= 18 + 2$$
$$= £20$$

The arithmetic mean of a frequency distribution

The arithmetic mean of a frequency distribution can be described also as the weighted arithmetic mean, or the arithmetic mean of grouped data. **This is calculated by multiplying each item by its frequency or weight, adding these up and dividing by the sum of the frequencies.**

The formula is:

$$\bar{x} = \frac{\Sigma fx}{\Sigma f}$$

where Σ is the sum of,

f is the frequency,

x is the value of the items.

With the use of the assumed mean this becomes:

$$\bar{x} = x \pm \frac{\Sigma fd_x}{\Sigma f}$$

where x is the assumed mean,

d_x is deviation from the assumed mean,

fd_x is the frequency times the deviation from the assumed mean.

For example, if 25 transistor radios stocked by a shop have prices (to the nearest £) as shown in Table 6.1, the arithmetic mean can be calculated in two ways.

If the assumed mean x is 28, the arithmetic mean can then be calculated as:

$$\bar{x} = x \pm \frac{\Sigma fd_x}{\Sigma f}$$
$$= 28 - \frac{50}{25}$$
$$= 28 - 2$$
$$= £26$$

Table 6.1 Frequency distribution and deviation from assumed mean.

Price (to the nearest £)	Number of transistor radios (f)	Deviation from assumed mean (d_x) $(x = 28)$	Frequency × deviation from assumed mean (fd_x)
20	2	−8	−16
24	6	−4	−24
25	10	−3	−30
30	4	+2	+8
32	3	+4	+12
	$\Sigma f = 25$		$\Sigma fd_x = -70 + 20 = -50$

This shows that the average price of the 25 transistor radios is £26. The method used to calculate the average was by using the assumed mean. This result can be checked by using the other method of calculation. This is to multiply the items by the frequencies and to divide the sum by the total frequencies (see Table 6.2 and the calculations that follow).

Table 6.2 Frequency distribution of prices of transistor radios.

Price (£) (x)	Frequency (f)	Price × frequency
20	2	40
24	6	144
25	10	250
30	4	120
32	3	96
	$\Sigma f = 25$	$\Sigma fx = 650$

The arithmetic mean can then be calculated as:

$$\bar{x} = \frac{\Sigma fx}{\Sigma f}$$

$$= \frac{650}{25}$$

$$= £26$$

This method helps to emphasise that, although there are some transistor radios priced at over £30 and others at £20, the average price for this selection is £26. If this was a random sample of all transistor radios then the consumer would know that £26 was likely to be a representative price for this type of radio (see Chapter 8).

The arithmetic mean of a grouped frequency distribution

The arithmetic mean of a grouped frequency distribution can also be described as the weighted arithmetic mean with frequency classes. The problem with class intervals is that there is no way of knowing the actual distribution of frequencies within a class. Therefore an assumption has to be made and this is usually that the frequencies equal the mid-point of the class interval (see Section 4.3).

There are two methods of calculating the arithmetic mean of a frequency distribution was class intervals:

- the mid-point method,
- the class interval method.

The mid-point method

This method uses the mid-points of each class to represent the classes in the calculation. The mid-point is usually found by adding together the lower and upper limits of the class and dividing by 2. With continuous data it can be assumed that class intervals such as '20 but less than 25' in fact mean '20 to 24.99'. The next class interval will be '25 to 29.99' and so on. The mid-point of the class interval '20 to 24.99' therefore is:

$$\frac{20 + 24.99}{2} = \frac{44.99}{2} = 22.49$$

Rounded to one decimal place this mid-point will be 22.5. It could be said that 22.5 is the 'common-sense' mid-point, in the sense that it does not suggest a higher degree of accuracy than is involved in the whole calculation.

Therefore for class intervals such as '20 but less than 25' the mid-point can be found simply by:

$$\frac{20 + 25}{2} = \frac{45}{2} = 22.5$$

With class intervals of '20 to 24' followed by '25 to 29', '30 to 34' and so on the mid-point can be found by:

$$\frac{20 + 24 + 1}{2} = \frac{45}{2} = 22.5$$

Instead of adding 1, it is possible to add the lower limit of the next class interval, $20 + 25$ (see Section 4.3 for a discussion of the problems of classification).

For example, the overtime pay for a number of employees has the grouped frequency distribution shown in Table 6.3.

If the assumed mean x is £37.5, the arithmetic mean can then be calculated as:

$$\bar{x} = x \pm \frac{\Sigma f d_x}{\Sigma f}$$

$$= £37.5 - \frac{145}{100}$$

$$= £37.5 - 1.45$$

$$= \mathbf{£36.05}$$

Table 6.3 Grouped frequency distribution and deviation from mid-point.

Overtime pay (£)	Number of employees (f)	Mid-points of class intervals	Deviation of mid-point from assumed mean (£37.5) (d_x)	Frequency × d_x (fd_x)	
20 but less than 25	11	22.5	−15	−165	
25 but less than 30	15	27.5	−10	−150	
30 but less than 35	16	32.5	−5	−80	
35 but less than 40	18	37.5	0	0	
40 but less than 45	30	42.5	+5		+150
45 but less than 50	10	47.5	+10		+100
	100			−395	+250
	$\Sigma f = 100$			$\Sigma fd_x = -145$	

The class-interval method

In this case instead of using the mid-points of the class intervals, the calculation is carried out using deviations in 'units of the class interval' (5). In Table 6.4 the deviations of classes from the class of the assumed mean are in units of 5, and this has to be allowed for in the calculation of the arithmetic mean by multiplying by 5.

If the assumed mean x is £37.5, the arithmetic mean can be calculated as:

$$\bar{x} = x \pm \frac{\Sigma fd_x}{\Sigma f} \times \text{class interval}$$

$$= £37.5 - \frac{29}{100} \times 5$$

$$= £37.5 - 0.29 \times 5$$

$$= £37.5 - 1.45$$

$$= \textbf{£36.05}$$

The class-interval method is very often the simpler method for calculation. However, if the class intervals are uneven, allowance must be made for this. For example, if the last class in Table 6.4 had been '45 but less than 55', then there would have been two units of 5 in this class. The deviation of this class from the class of the assumed mean would have to be +2.5. If the mid-point of this larger class is taken as 50, then this is 2.5 units of 5 away from the assumed mean of 37.5. With a number of unequal class intervals it may be easier to use the mid-point method.

Table 6.4 Grouped frequency distribution and deviation of classes from the class of the assumed mean.

Overtime pay (£)	Number of employees (f)	Deviation of classes from class of the assumed mean (d_x)	Frequency × (fd_x)
20 but less than 25	11	−3	−33
25 but less than 30	15	−2	−30
30 but less than 35	16	−1	−16
35 but less than 40	18	0	0
40 but less than 45	30	+1	+30
45 but less than 50	10	+2	+20
	100		−79 +50
			$\Sigma fd_x = -29$

Advantages and disadvantages of the arithmetic mean

The arithmetic mean involves the use of all the data and all the values and this major strength is also a weakness under certain circumstances. The arithmetic mean of 2, 5, 6, 8 and 129 is 30. This answer is not close to any of the actual values, because all the values are used, including the extreme ones. However, because the arithmetic mean can be calculated precisely, statisticians prefer it to other averages for most purposes.

Advantages of the arithmetic mean
- It is widely understood and the basic calculation is straightforward.
- It makes use of all the data in the group and it can be determined with mathematical precision.
- It can be determined when only the total value and the number of items are known.

Disadvantages of the arithmetic mean
- A few items of a very high or very low value may make the mean appear unrepresentative of the distribution.
- It may not correspond to an actual value and this may make it appear unrealistic.
- When there are open-ended class intervals, assumptions have to be made which may not be accurate.

6.3 The median

The median can be defined as the value of the middle item of a distribution which is set out in order.

Example 1
If five people earn

  £60, £70, £100, £115 and £320,

the median is the value of the middle item: £100.

In a discrete distribution such as this one, with an odd number of items, the median can be ascertained by inspection.

The formula for finding the median position for a discrete series is:

$$\text{the median position} = \frac{n+1}{2}$$

where n is the number of items.

In the example above, the median position is:

$$\frac{5+1}{2} = 3$$

So the median is the value of the third item, £100.

If there are an even number of items, the median is found by adding the two middle items together and dividing by 2. The median position is found in the same way as before.

Example 2
If six people earn

  £60, £70, £100, £110, £115, £320,

$$\text{the median position} = \frac{n+1}{2}$$

$$= \frac{6+1}{2}$$

$$= 3.5$$

$$\text{the median} = \frac{£100 + £110}{2}$$

$$= \textbf{£105}$$

The symbol for the median is M. In this example the position of M is between the third and fourth item and therefore the value of M is £100 plus £110 divided by 2: £105.

This example shows that **the median is unaffected by extreme values** (such as the £320) which is one reason for using the median as an average in certain circumstances. It is often used with wage distributions.

Also, **the median divides a distribution in half by the number of items** (not their values). In the examples above, where the median is £100 there are two items on either side, and where the median is £105 there are three items on either side.

The median of a grouped frequency distribution

With a *grouped frequency distribution* the median can be found by two methods:

- calculation,
- graphically.

The median found by calculation

The position of the median for a continuous series is $n/2$ or $f/2$, where n is the number of items and f is the frequency.

Table 6.5 shows the grouped frequency distribution of overtime pay for 50 employees, so:

$$\text{the median position} = \frac{50}{2} = 25$$

The 25th employee's overtime pay is therefore the median pay. This falls within the class interval '£15 but less than £20'. Therefore the 25th employee earns at least £15 in overtime pay, but less than £20.

Table 6.5   Overtime pay: grouped frequency distribution.

Overtime pay (£)	Number of employees	Cumulative frequency
less than 5	3	3 people received less than £5
5 but less than 10	5	8 people received less than £10
10 but less than 15	14	22 people received less than £15
15 but less than 20	12	34 people received less than £20
20 but less than 25	10	44 people received less than £25
25 but less than 30	6	50 people received less than £30
	50	

The only way to arrive at a closer calculation of his or her pay is to assume that the earnings are evenly distributed within the class interval. The 25th employee is the third person in the group, because the previous group includes the 22nd employee and $25 - 22 = 3$.

If the class interval value (£5) is divided evenly between the 12 people in the group, and the 25th employee is the third person of these 12, then

$$\text{the median} = £15 + \frac{3}{12} \times 5$$

$$= £16.25$$

where 15 is the lower class interval of the class in which the median falls, and $\frac{3}{12} \times 5$ indicates that the median pay is the third out of 12 sharing £5. (Notice that it is only the class interval of the class in which the median falls that is important; therefore irregular class intervals in a distribution are not a problem because they can be ignored.)

In this distribution the median overtime pay is the earnings of the 25th employee which is £16.25.

This divides the distribution in half, so that of the 50 employees half will earn less than £16.25 and half will earn more. In this sense the median provides a good representation of the grouped frequency distribution.

The median found graphically

The median of a grouped frequency distribution can also be found by drawing the cumulative frequency curve or ogive. **'Ogive' is an architectural term which is used to describe an S (or flattened S) shape,** similar to the usual shape of the cumulative frequency curve.

Using the information in Table 6.5, the graph in Figure 6.1 can be drawn, plotting the cumulative frequency against overtime pay. Notice that, when plotting the frequencies, the point must be placed at the end of the group intervals (5, 10, 15, etc.).

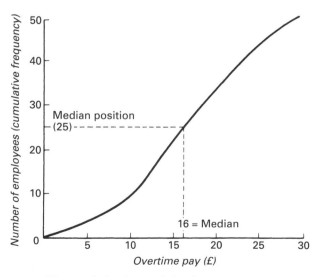

Figure 6.1 A cumulative frequency curve

The median position $= \dfrac{50}{2} = 25$

A horizontal line is drawn from this point (25) to the ogive, and at the point of intersection a vertical line is dropped to the horizontal axis. This shows that the median is about £16. The more accurately the graph is drawn, the more accurately it is possible to interpolate the median.

Advantages of the median
- Extreme high and low values do not distort it as a representative average. Therefore it is useful for describing distributions in areas such as wages where a few extreme values would distort the arithmetic mean.
- It is straightforward to calculate even if not all the values are known, or where there are irregular class intervals.
- It is often an actual value and even when it is not it may 'look' representative and realistic.

Disadvantages of the median
- It gives the value of only one item. The other items are important in ascertaining the position of the median, but their values do not influence the value of the median. If the values are spread erratically, the median may not be a very representative figure.
- In a continuous series, grouped in class intervals, the value of the median is only an estimate based on the assumption that the values of the items in a class are distributed evenly within the class.
- It cannot be used to determine the value of all the items; the number of items multiplied by the median will not give the total for the data; therefore it is not suitable for further arithmetical calculations.

6.4 The quartiles

The median divides an ordered distribution into half, and in a similar way the quartiles divide it into four equal parts. It is also possible to divide distributions into tenths, and so on, but the most frequently used division is into quartiles.

There are 'three' quartiles:

- the lower quartile Q_1,
- the middle quartile Q_2 (this is the same as the median M),
- the upper quartile Q_3.

In a distribution of 100 items the lower and upper quartiles will be the values of the 25th (Q_1) and the 75th (Q_3) items. With the median, which will be the value of the 50th item, the quartiles will divide this distribution into four equal parts: 1–25, 26–50, 51–75, 76–100.

The method of calculating the lower and upper quartiles is very similar to the method for calculating the median.

Table 6.6 Overtime pay: grouped frequency distribution.

Overtime pay (£)	Number of employees	Cumulative frequency
less than 5	3	3
5 but less than 10	5	8
10 but less than 15	14	22
15 but less than 20	12	34
20 but less than 25	10	44
25 but less than 30	6	50
	50	

The position of the lower quartile (Q_1) for a continuous series is $n/4$. Using the data in Table 6.6, this will be $50/4 = 12.5$.

The pay received by a theoretical employee lying between the 12th and 13th is the lower quartile pay.

This lies in the class '£10 but less than £15', which is shared by 14 employees. Therefore:

$$Q_1 = £10 + \frac{4.5}{14} \times 5$$

$$= £10 + \frac{22.5}{14}$$

$$= £10 + 1.607$$

$$= \mathbf{£11.61}$$

The position of the upper quartile (Q_3) = $3n/4$. Using the data in Table 6.6, this will be $3 \times 50/4 = 150/4 = 37.5$.

The pay received by a theoretical employee lying between the 37th and 38th employee is the upper quartile pay.

This lies in the class '£20 but less than £25', which is shared by 10 employees.

$$Q_3 = £20 + \frac{3.5}{10} \times 5$$

$$= £20 + \frac{17.5}{10}$$

$$= £20 + 1.75$$

$$= \mathbf{£21.75}$$

Therefore this distribution is divided in the following way:

Q_1 = 12.5th employee receiving £11.61 in overtime pay
M = 25th employee receiving £16.25 in overtime pay
Q_3 = 37.5th employee receiving £21.75 in overtime pay

Graphically this can be interpolated as shown in Figure 6.2.

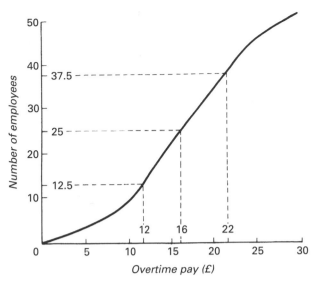

Figure 6.2 The lower and upper quartiles

From the division of the distribution in Table 6.5 and Figure 6.2 by the median and the lower and upper quartiles, it can be stated that:

- half the employees are earning less than £16.25 in overtime pay, and half are earning more than this;
- half the employees are earning between £11.61 and £21.75 (that is, between Q_1 and Q_3) in overtime pay;
- a quarter of the employees are earning less than £11.61 in overtime pay, and a quarter more than £21.75.

The difference between the upper and lower quartile is called the inter-quartile range or the quartile deviation (see Section 7.3).

In the same way as the median, the quartiles are useful for descriptive purposes. The results produced from the distribution in Table 6.5 could be compared with those from a similar distribution based on another company.

Box and whisker diagram

A box (or box plot) diagram is used to display the values of the smallest observation, the largest observation, the median and the quartiles of a set of data. Using the data in Table 6.5 and the calculations above, for example, the median is £16.25, the lower quartile is £11.61, the upper quartile is £21.75, the smallest observation is £0 and the largest £30.00. This produces the diagram shown in Figure 6.3.

A scale line is used for the variable (overtime pay). The rectangular box has its vertical sides located at the lower and upper quartiles, and the vertical line inside the box at the median. The dots indicate the location of the smallest and largest observations.

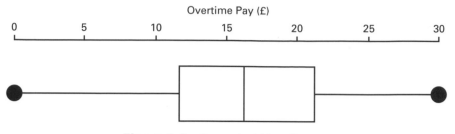

Figure 6.3 Box and whisker diagram

The horizontal lines from the box to the dots can be referred to as 'whiskers'. In Figure 6.3 the whisker on the left is slightly longer than the one on the right, indicating that the spread of the smaller observations is greater than that of the larger observations. The median is closer to the lower quartile than to the upper quartile, which shows that the distribution is positively skewed.

6.5 The mode

The mode can be defined as the most frequently occurring value in a distribution. It is an average that is frequently used in conversation when reference is made to such things as an 'average' income, an 'average' person and so on. It may be stated that 'the average family has two children', meaning that 'most families' have two children. The arithmetic mean may show that the average family has 2.3 children, but although this may be mathematically correct it does not appear very sensible. In such cases as this, the mode may be a more 'sensible' average to use.

> *Example 1*
> The mode of the figures 4, 4, 5, 6, 11 is 4, because it is the most frequently occurring number.

In a frequency distribution the mode is the item with the highest frequency.
Calculating the mode for a grouped frequency distribution is not easy because since a grouped frequency distribution does not have individual values it is impossible to determine which value occurs most frequently. It is possible to calculate the mode, but it is not particularly useful to do so. **For most purposes the modal class of a grouped frequency distribution is perfectly satisfactory as a form of description.**

The modal class is the one with the highest frequency. The problem with using it is that, if a different set of class intervals had been chosen when the classifications were being decided, the modal class would have included different values. However, it can be useful as a form of description.

Table 6.7 shows again the grouped frequency distribution of overtime pay for 50 employees.

Example 2

Size of component (cm)	Number of makes of car using the size
20	4
21	10
22	15
23	20
24	1

The mode in this example is 23 cm, because 20 is the highest frequency; 23 cm is the modal size of component or the most frequently used size.

The modal class is '£10 but less than £15' because this class has the highest frequency (14). If the classification and frequencies had been:

0 but less than 10	8
10 but less than 20	22
20 but less than 30	16

then the modal class would have been '£10 but less than £20'. Although this would have produced a different modal class from the same distribution, the modal class would still have been around the centre because this is a unimodal distribution (only one modal class) which is not very skewed (see Section 7.1).

The modal class can be illustrated by drawing a histogram (Figure 6.4). From the histogram it is possible to estimate the mode. This estimation is carried out by drawing a line from the top right-hand corner of the modal class rectangle or block to the point where the top of the next adjacent rectangle to the left meets it (line *A* to *B* in Figure 6.4); and a corresponding line from the left-hand top corner of the modal class rectangle to the top of the class on the right (line *C* to *D*). Where these two lines cross, a vertical line can be dropped to the

Table 6.7 Overtime pay: grouped frequency distribution.

Overtime pay (£)	Number of employees
less than 5	3
5 but less than 10	5
10 but less than 15	14
15 but less than 20	12
20 but less than 25	10
25 but less than 30	6
	50

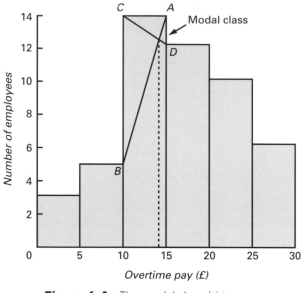

Figure 6.4 The modal class: histogram

horizontal axis and this will show the value of the mode (about £14 in Figure 6.4). This can only be an estimate, because the individual values are not known. For most purposes, however, the modal class provides a sufficient description of this aspect of a grouped frequency distribution.

Some distributions have more than one modal class, because two or more classes have the same frequency. These distributions are called bi-modal, tri-modal and so on. Figure 6.5 shows a bi-modal distribution.

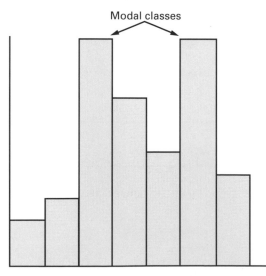

Figure 6.5 A bi-modal distribution

Advantages of the mode

- It is a commonly used average, although people do not always realise that they are using it.
- It can be the best representative of the typical item, because it is the value that occurs most frequently.
- It has practical uses. For instance, employers will often adopt the modal rates of pay, the rate paid by most other employers. Cars and clothes are made to modal sizes, houses are built on the basis of the modal average size of the family.
- Modal information can often be supplied quickly by people who have experience in a particular area.
- The mode is often an actual value and therefore may appear to be realistic and 'sensible'.

Disadvantages of the mode

- It may not be well defined and can often be a matter of judgement.
- It does not include all the values in the distribution.
- It is not very useful if the distribution is widely dispersed.
- It is unsuitable for further or other kinds of calculation because of its lack of exactness.

6.6 The geometric mean

The geometric mean is defined as the nth root of the product of the distribution (where n is the number of items in the distribution). If there are three items it is the third root (cube root), and so on. **This average is used to measure changes in the rate of growth** (see Section A3.14 on the geometric progression).

THE GEOMETRIC MEAN

Example 1
The geometric mean of 3, 4, 15 is:

$$\sqrt[3]{3 \times 4 \times 15} = \textbf{5.65}$$

Example 2
If the price of commodity A has risen from £60 to £120, this is an increase of 100%.

If the price of commodity B has risen from £80 to £100, this is an increase of 25%.

The arithmetic mean of these percentage increases would be:

$$\frac{100\% + 25\%}{2} = \textbf{62.5\%}$$

The geometric mean would be:

$$\sqrt{100 \times 25} = \textbf{50\%}$$

Whether the 'true' increase in Example 2 is 62.5% or 50% is open to argument (see Section 9.1 for a discussion on this in relation to index numbers), however the rate of growth can be described in other terms and the geometric mean is seldom used.

6.8 The harmonic mean

The harmonic mean is used to average rates rather than simple values (averaging kilometres per hour, for example). It is rarely used except in engineering.

ASSIGNMENTS

1 *Profits of three companies*

	Company A (£)	Company B (£)	Company C (£)
Year 1	2 000	13 000	10 000
Year 2	4 000	2 000	9 000
Year 3	6 000	6 000	8 000
Year 4	9 000	2 000	6 000
Year 5	14 000	12 000	2 000

Calculate the annual average (arithmetic mean) profit for these three companies.

Comment on the results and on the profit figures.

What advice would be reasonable to give to someone thinking of buying one of these companies?

2 *Median and quartiles*
A company employing 100 part-time employees makes the following monthly payments:

Part-time pay (£)	Number of people	Cumulative totals
20 and less than 25	5	5
25 and less than 30	8	13
30 and less than 35	17	30
35 and less than 40	39	69
40 and less than 45	14	83
45 and less than 50	12	95
50 and less than 55	5	100
	100	

(i) From the data calculate the median and quartiles.

(ii) Draw an ogive and interpret the median and quartile. Comment on the wage distribution of this company.

3   In a factory the operation of four machines is recorded. The results are as follows:

Machine	Day 1	Units of output Day 2	Day 3	Day 4	Day 5
A	500	501	505	504	494
B	560	558	572	560	562
C	530	475	538	442	520
D	460	458	452	462	430

The modal output for all the machines of this type is 500 units a day.
Comment on the performance of these four machines.

4   Discuss the purpose of calculating averages. Why is there more than one type of average?

5   Find references to the arithmetic mean and the median in government publications. Comment on the way these averages are used in these cases.

7 Summarising data: dispersion

OBJECTIVES

- To analyse the role of measures of dispersion
- To provide an understanding of the calculation of the range, the interquartile range, the standard deviation and the variance
- To consider the advantages and disadvantages of each measure of dispersion

7.1 Dispersion

Data can be summarised and compared by measures of dispersion (which are also described as measures of deviation or spread) just as data can be summarised and compared by the use of averages.

If items are widely dispersed, averages do not provide a clear summary of a distribution; they do not give an indication of the form or shape of a distribution. **Distributions are not only clustered around a central point, but also spread out around it.**

The distribution shown in Figure 7.1 is of a bell-shaped or normal curve type (see Section 8.3 for more on the normal curve). It is symmetrical in that the distribution is spread equally on either side of the arithmetic mean, median and mode.

Some frequency distributions are 'skewed', so that the peak is displaced to the left or right (Figures 7.2 and 7.3).

When the peak is displaced to the left of centre, the distribution is described as being positively skewed when the peak is displaced to the right of centre, the distribution is described as negatively skewed. With these distributions the mode is located at the highest point while the median usually lies between the mode and the mean.

Other commonly occurring shapes of distributions are the bi-modal, the rectangular and the *J*-shaped (Figures 7.4, 7.5 and 7.6).

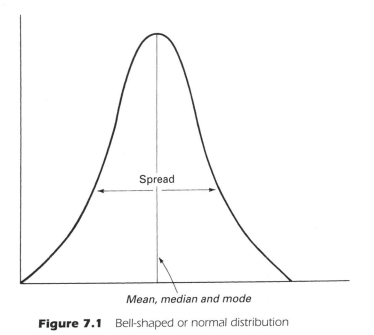

Spread

Mean, median and mode

Figure 7.1 Bell-shaped or normal distribution

As a method of providing a short description of distributions of data an average may not be sufficient because it provides an indication of the central value and no more. **In many circumstances what is needed is a measure of dispersion which provides an indication of the deviation of the data around this central value.**

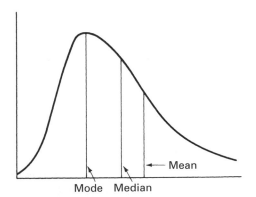

Mode   Median   ← Mean

Figure 7.2 Positively skewed distribution

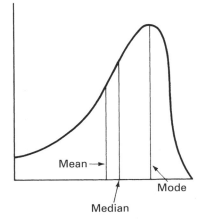

Mean →

Mode

Median

Figure 7.3 Negatively skewed distribution

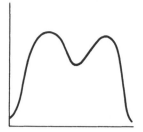

Figure 7.4 Bi-modal distribution

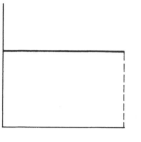

Figure 7.5 Rectangular distribution

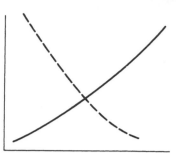

Figure 7.6 J-shaped and reverse J-shaped distributions

DISPERSION

For both the following sets of data the arithmetic mean is 32:

    Firm *A*   3, 4, 7, 16, 20, 30, 38, 53, 61, 88
    Firm *B*   20, 21, 23, 28, 31, 34, 37, 39, 49, 50

In both cases

$$\bar{x} = \frac{320}{10} = 32$$

If a firm sells a variety of makes of a product at an average price of £32, this does not give an indication of the price of the cheapest make it sells or the price of the most expensive. In other words, it does not give an idea of dispersion. If the sets of scores above represent the prices charged by two firms for makes of a particular product they have in stock, then although their average price is the same (£32) the first firm carried very much cheaper (£3) and very much more expensive (£88) makes or brands of this product than the second firm (£20 and £50).

In other words, in the above example, firm *A* has a wider price range (or its prices are more dispersed or more spread out) for this particular product than firm *B* has. This can be further illustrated by a graph (Figure 7.7).

It is possible to calculate a measure of spread for these two distributions by using the simplest form of dispersion, the range.

7.2 The range

The range can be defined as the highest value in a distribution minus the lowest.
The range is an everyday method of describing the dispersion or spread of data.

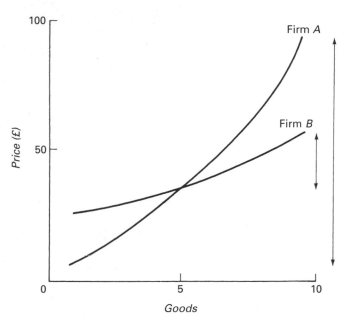

Figure 7.7 Price distribution: dispersion

RANGE

In the distributions in Figure 7.7 the ranges are:

Firm *A*   £88 − £3 = £85
Firm *B*   £50 − £20 = £30

Therefore it is possible to say that firm *A* has makes or brands of this product which have an average price of £32 with a range of £85, while firm *B* has makes or brands of the same product with an average price of £32 and a range of £30. The larger the range the wider the dispersion or spread of the distribution. Therefore firm *A* has a wider spread of prices than firm *B*.

The range is often used with the mode. In a shop it is possible to ask (about, for example, watches): 'What price range do you have?'. The answer might be: 'They range from a price of £5 to £200.' This is one simple statement which provides the dispersion or spread of prices for this commodity. The next question could be: 'What is the average price?'. After all, the fact that the watches cost between £5 and £200, does not give any indication whether most of them are priced close to £5 or close to £200, or spread evenly through the range. The answer might be: 'the usual price is about £20', or 'most of our customers buy the ones costing around £20', or 'the £20 ones are the most popular'. Notice that all three of these answers use the mode rather than the arithmetic mean.

It is now clear that the price of these commodities ranges from £5 to £200 with a modal price of about £20. This means that a customer who buys a watch for £100 is a rare customer buying an expensive watch. The customer who buys a watch for £5 will be buying a very cheap one.

The range and an average can provide a very clear summary of the variations in the prices (or weights, lengths or sizes) of goods available. This summary does not depend on the number of articles. In the example above, there could have been 20 watches or 200 or 2000.

All measures of dispersion are designed to provide similar information to the range, but information which is either more precise or expressed in a different way.

7.3 The interquartile range

The interquartile range (or quartile deviation) can be defined as the difference between the upper quartile and the lower quartile ($Q_3 - Q_1$). (As was seen in Section 6.5, a distribution can be divided into four equal parts and the point marking each quarter is called a quartile.)

Table 7.1 shows the overtime wage list of a company employing 20 people. The wages of employees 5, 10 and 15 on this list divide the distribution into four equal parts (1–5, 6–10, 11–15, 16–20). The position of the lower quartile, Q_1, is:

$$\frac{n}{4} = \frac{20}{4} = 5$$

The wage of the fifth employee is the lower quartile wage: **£50**.

The position of the upper quartile, Q_3, is:

$$\frac{3n}{4} = \frac{60}{4} = 15$$

The wage of the fifteenth employee is the upper quartile wage: **£110**.

$$\begin{aligned}
\text{The interquartile range} &= Q_3 - Q_1 \\
&= £110 - £50 = \textbf{£60}
\end{aligned}$$

Table 7.1 Overtime pay of 20 employees.

Employees	Overtime pay (£)	Employees	Overtime pay (£)
1	40	11	89
2	42	12	93
3	43	13	97
4	48	14	100
5	50	15	110
6	60	16	140
7	62	17	200
8	65	18	210
9	71	19	212
10	80	20	220

It is also possible to arrive at the **semi-interquartile range** by dividing the interquartile range by 2.

$$\text{The semi-interquartile range} = \frac{Q_3 - Q_1}{2}$$

$$= \frac{£110 - £50}{2}$$

$$= \frac{£60}{2} = \textbf{£30}$$

The interquartile range helps to summarise and clarify the distribution. For the data in Table 7.1 the interquartile range (£110 minus £50) covers those employees earning the middle range of overtime pay, and this includes 50% of the employees working in the company.

The interquartile range of one company can be compared with that of another. For instance, a company with an interquartile range of rates of overtime pay of £60 has a more widely dispersed middle range of overtime pay than a company with an interquartile range of £20.

Another method of arriving at the value of the interquartile range is by using the ogive (see Section 6.4). Using the data from Table 7.2, the ogive can be drawn as in Figure 7.8.

$$Q_1 = £12$$
$$M = £16$$
$$Q_3 = £22$$
$$Q_3 - Q_1 = £22 - £12 = \textbf{£10}$$

The interquartile range is £10 in Figure 7.8, and therefore the middle 50% of the employees of the firm earn a median overtime pay of £16 with an interquartile dispersion of £10.

This result could be compared with that of another firm with a similar median overtime pay among its employees but with an interquartile range of £20. This indicates that in this second company the middle range of employees could earn both less and more overtime pay than the middle range of employees

Table 7.2 Overtime pay for 20 employees.

Overtime pay (£)	Number of employees	Cumulative frequency
less than 5	3	3
5 but less than 10	5	8
10 but less than 15	14	22
15 but less than 20	12	34
20 but less than 25	10	44
25 but less than 30	6	50
	50	

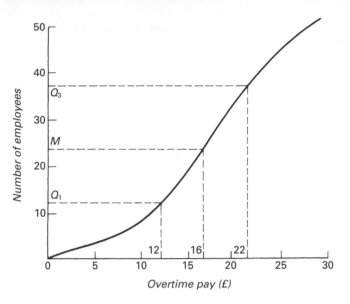

Figure 7.8 Interquartile range

in the first company. Instead of the interquartile range being £22 − £12, it could be say £27 − £7.

Notice that **the interquartile range is not influenced by extreme items; so that very small and very large values do not alter its general spread.** For example, if the owner of a company draws an annual salary of £100 000 and the works manager receives an annual salary of £20 000 while the rest of the company employees receive wages between £4000 and £8000 per year, the very high wages will not 'distort' the interquartile range. The 'simple' range would include the extreme figures (range: £4000–£100 000) and the result does not give a very clear picture of the general level of pay in the company; therefore in these circumstances the interquartile range is preferable.

7.4 The standard deviation

The standard deviation as a measure of dispersion

The standard deviation is a measure of dispersion which uses all the values in a distribution in the sense that every value contributes to the final result in the same way that every value contributes to the calculation of the arithmetic mean (see Section 6.2). Neither the range nor the interquartile range make use of all the scores or values in a distribution. The range relies on the highest and lowest values, the interquartile range on the values of the upper and lower quartiles. Therefore both of them rely on the position of two points in a distribution and the values of the other items are not very important. This can lead to 'distortion' in the sense that it is possible that the two values are not very representative of all the other values in the distribution.

The standard deviation is the 'standard' measure of dispersion both because it is standardised for all values of n, and because it is very useful both practically and mathematically. It is important because of its mathematical properties and use in sampling theory (see Chapter 8) rather than because it makes a distribution more easily understood. It is therefore basically different from the range and the interquartile range which can quickly make a distribution more easily understood but which are of limited use mathematically.

The standard deviation can be used as a measure of dispersion in all symmetrical and unimodal distributions, and also in distributions that are moderately skewed (see Section 7.1). These forms of distribution are frequently found in sampling and in surveys in many areas, for instance in examination results and in quality control.

The standard deviation shows the dispersion of values around the arithmetic mean. The greater the dispersion the larger the standard deviation. Therefore a distribution with an arithmetic mean of £10 and a standard deviation of £4 has a greater or wider dispersion than a similar distribution with a standard deviation of £2. This can be shown graphically by saying that curve A (in Figure 7.9) has a wider spread or dispersion than curve B. They both have the same number of items and share the same arithmetic mean.

In Figure 7.9 both curves are of the 'normal curve' type. The further away a distribution is from this type the more difficult it becomes to interpret the standard deviation accurately.

With a normal curve it has been calculated that it is possible to mark off the area under the curve into certain proportions. This can be seen in Figure 7.10. In this curve the arithmetic mean is shown in the centre with three standard deviations marked off on either side. It has been calculated that approximately 68% of the items of a distribution will lie within one standard deviation on either side of the arithmetic mean (two standard deviations in all).

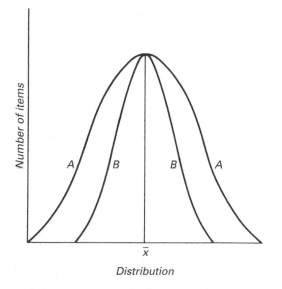

Distribution

Figure 7.9 Normal distributions with different dispersions

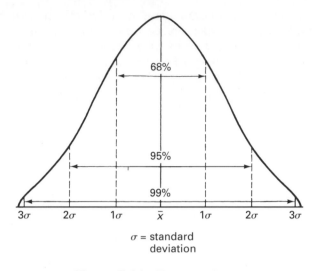

Figure 7.10   The normal curve

Approximately 95% of the items will lie within two standard deviations (or 1.96 standard deviations) on either side of the mean; and approximately 99% (in fact 99.74%) of the items will lie within three standard deviations on either side of the mean.

These are approximations, but this means that it is possible to arrive at a fairly clear picture of a distribution if the arithmetic mean, the standard deviation and the number of items are known.

STANDARD DEVIATION AND DISTRIBUTION

With a mean of £20, a standard deviation of £5, and 500 items in the distribution, it is possible to draw the curve in Figure 7.11. This curve is

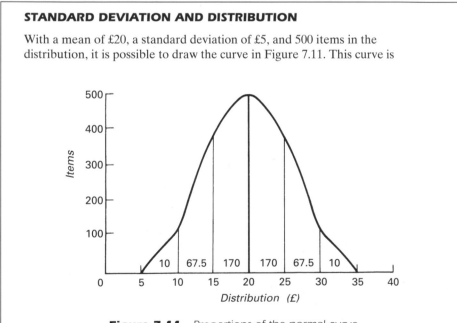

Figure 7.11   Proportions of the normal curve

constructed by putting in the mean at the centre of the horizontal axis (£20), drawing in the vertical to represent the number of items (500) and then marking the horizontal scale by adding and subtracting the standard deviation (£5) from the mean (£20 ± 5 for one standard deviation, £20 ± 10 for two standard deviations and £20 ± 15 for three standard deviations). The shape of the curve can then be arrived at approximately by assuming that 68% of the items lie within one standard deviation on either side of the mean (say 340 out of 500 items); 95% of the items lie within two standard deviations of the mean (a total of approximately 475 out of 500 items); and nearly 100% of the items (495 out of 500) lie within three standard deviations of the mean. Strictly, the number of items under each section of the curve should be proportional to the area of that section.

Calculation of the standard deviation

To calculate the standard deviation, the method used is to take the deviation (the difference) of each value from an average and then calculate an average from these deviations for the number of cases involved. The problem is that if the actual deviations from the arithmetic mean are added together the result will always be zero, because the positive and negative deviations will cancel each other out.

DEVIATIONS

For the values 5, 6, 8, 11, 15, the arithmetic mean is 9 (45 ÷ 5). The deviation of each value from 9 is shown below.

Value	Deviation	
5	−4	
6	−3	
8	−1	
11		+2
15		+6
	−8	+8

In the above example, the positive and negative deviations cancel each other out. Therefore, to obtain a measure of dispersion around the mean, the negative signs must be removed. There are two methods of doing this.

1 By ignoring the signs and taking the absolute values of the deviations (16 in the above example). This method is used to arrive at the **mean deviation**. The mean deviation is the arithmetic mean of the absolute differences of each value from the mean (16 ÷ 5 = 3.2). The mean deviation is seldom used, because it is not mathematically very useful compared with the standard deviation.
2 By squaring the deviation. This has the effect of making the negative

deviations positive. The square root of the result is taken to cancel out the squaring of the deviations. This is the method used to calculate the standard deviation.

Therefore **the standard deviation is calculated by adding the square of the deviations of the individual values from the mean of the distribution, dividing this sum by the number of items in the distribution and finding the square root of the result.** Taken step by step this is less complicated than it sounds:

Step 1. The first step in calculating the standard deviation is to calculate the arithmetic mean by using the method of deviations from an assumed mean (as in Section 6.3). For example, a survey of the weekly expenditure on transport for five families gives the results shown in Table 7.3.

Table 7.3 Transport expenditure and deviations from assumed mean.

Families	Weekly expenditure on transport (£)	Deviation from the assumed mean (£25)	
1	15	−10	
2	25	0	
3	26		+1
4	20	−5	
5	14	−11	
		−26	+1 = −25

From Table 7.3, over all five values the difference between the assumed mean and the real (arithmetic) mean is −£25. Therefore the average deviation is:

$$£\frac{-25}{5} = -5$$

and the arithmetic mean is:

$$£25 - 5 = £20$$

Step 2. The second step is to square the deviation of the value of each item from the arithmetic mean (Table 7.4). In Table 7.4 the sum of the squared deviations (the sum of the squares) is £122.

Step 3. This sum (£122) is divided by the number of items in the distribution:

$$£\frac{122}{5} = £24.4$$

Step 4. The final step is to find the square root of this figure:

$$\sqrt{24.4} = £4.9 \text{ approximately.}$$

Therefore the standard deviation of this distribution is £4.9 and the arithmetic mean is £20.

Table 7.4 Deviations from arithmetic mean, and squared deviations.

Families	Weekly expenditure on transport (£)	Deviation from arithmetic mean (20) (d_x)		Deviation squared (d_x^2)
1	15	−5		25
2	25		+5	25
3	26		+6	36
4	20		0	0
5	14	−6		36
				122

The basic formula for calculating the standard deviation is:

$$\sigma = \sqrt{\frac{\Sigma(d_x)^2}{n}}$$

where σ represents the standard deviation (sometimes S or S.D. is used),
$\Sigma(d_x)^2$ represents the squared deviations from the mean,
n is the number of items (f in a frequency distribution).

The standard deviation is sometimes written:

$$\sigma = \frac{\Sigma(x - \bar{x})^2}{n}$$

where x is the value of the item and $\bar{x}$ is the mean.

The distribution in Table 7.4 is unimodal and fairly symmetrical. By drawing the histogram and polygon it is clear that it is partly skewed (Figure 7.12). The fact that this distribution is partly skewed is reflected by the distribution of items. While 3.4 of the items represent 68% of the total number of items (i.e. families), they cover approximately 75% of the values (the three central values: 25, 26 and 20 and say four from the other two values – making 75 out of 100). Also, 4.8 of the items, representing 95% of the total number of items in the distribution, cover approximately 98% of the values.

Looking at the standard deviation in this way makes it clear whether the calculation is reasonable or not. To put it another way, one standard deviation on either side of the mean should cover about 68% of the items in the distribution as has been seen above. In fact in this example the range £15 to £25 (£20 plus and minus £5) includes the values of three out of five items. This is 60%, which is very close to 68%, given the small number of items involved and the fact that the distribution is skewed to some extent. This indicates that the standard deviation calculated as £4.9 for this distribution is reasonable.

If the standard deviation in the example had been calculated as £2, the limits around the arithmetic mean (£20) for one standard deviation would have

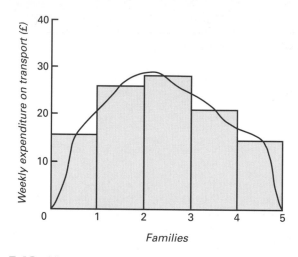

Figure 7.12 Histogram and polygon for transport expenditure survey

been £18 to £22. This would have included the value of only one item (family number 4 with a weekly expenditure on transport of £20), and therefore the standard deviation would appear to be too small.

On the other hand, if the standard deviation had been calculated as £8, the limits around the arithmetic mean for one standard deviation would have been in the range £12 to £28. This range would include all the values in the distribution, and therefore the standard deviation would appear too large.

This summarisation of a distribution is particularly important where there are a large number of items and where there is a grouped frequency distribution. Computer programs often produce either the standard deviation as a summary figure, or a figure which is based on the standard deviation (such as the variance and the coefficient of variability: see Sections 7.5 and 7.6).

Calculation of the standard deviation of a grouped frequency distribution

This is more complicated than the example used so far, but the overall method is similar and the interpretation is the same.

The method of calculation used here is similar to that used to find the arithmetic mean for a group distribution (see Section 6.2). It would be possible to use deviations from the mid-points of the classes, but to avoid using unwieldy figures in the last two columns it is easier to use units of class intervals (particularly with even class intervals). These 'class-interval' units have to be reconverted to the units of measurement at the end.

Table 7.5 shows the simplest method of calculation and makes it possible to calculate the arithmetic mean and the standard deviation from the same table. It gives the results of a survey into the expenditure on transport each month by

100 families. Σfd_x^2 is the sum of: the frequencies $\times$ (deviations from the assumed mean squared). It is found, in Table 7.4, by multiplying columns 3 and 4, but it could be found by squaring column 3 ($-4^2 = 16$) and multiplying the result by column 2 ($16 \times 2 = 32$).

In Table 7.5 the calculations are based on the assumed mean (£45). The arithmetic mean and standard deviation are:

$$\bar{x} = 45 + \frac{21}{100} \times 10 = 45 + 2.1 = \textbf{£47.10}$$

$$\sigma = \sqrt{\frac{\Sigma fd_x^2}{\Sigma f} - \left(\frac{\Sigma fd_x}{\Sigma f}\right)^2} \times \text{class interval}$$

$$= \sqrt{\frac{341}{100} - \left(\frac{21}{100}\right)^2} \times 10$$

$$= \sqrt{3.41 - 0.0441} \times 10$$

$$= \sqrt{3.3659} \times 10$$

$$= 1.835 \times 10 = \textbf{£18.35}$$

The arithmetic mean of £47.10 and the standard deviation of £18.35 can be used to compare the dispersion in this data with other data, provided the distributions are of the same kind and are measured in the same units.

Notice that one standard deviation on either side of the mean ($47 - 18 = 29$, $47 + 18 = 65$) gives limits of about £29 and £65. These limits include the frequencies 15, 18, 20 and half of 17 (say 8). This is a total of 61 out of 100, which is close to 68%.

Two standard deviations on either side of the arithmetic mean gives limits of about £11 and £83. These limits include all the frequencies except the first and last and therefore about 96% of the distribution. This is very close to 95%.

In fact the distribution is not perfectly symmetrical and this accounts for the difference between 61% and 68%, and 96% and 95%. However, it is clear that the result arrived at for the standard deviation of this distribution is reasonable.

Table 7.5 Data to be used for finding the standard deviation of a grouped frequency distribution.

(1) Expenditure (in classes of 10) (£)	(2) Number of households (f)	(3) Deviation of classes from class of assumed mean (d_x)	(4) (2) × (3) Frequency × deviations from assumed mean (fd_x)	(5) (3) × (4) Frequency × deviation from assumed mean, squared (fd_x^2)
under 10	2	−4	−8	32
10 and under 20	6	−3	−18	54
20 and under 30	12	−2	−24	48
30 and under 40	15	−1	−15	15
40 and under 50	18	0	0	0
50 and under 60	20	+1	+20	20
60 and under 70	17	+2	+34	68
70 and under 80	8	+3	+24	72
80 and under 90	2	+4	+8	32
	100		; +86	341
	$\Sigma f = 100$		= +21	$\Sigma fd_x^2 = 341$
			$\Sigma fd_x = +21$	

7.5 The variance

The variance is the average of the square of the deviations.

CALCULATING THE VARIANCE

Values	Deviation from arithmetic mean (11)	Deviation squared
5	−6	36
6	−5	25
10	−1	1
14	+3	9
20	+9	81
		152

$$\text{Variance} = \frac{152}{5} = 30.4$$

The variance is calculated in the same way as the standard deviation, but without the final step of finding the square root. The standard deviation for the distribution in Table 7.4 would be approximately 5.5. So **the variance can be interpreted as the square of the standard deviation (σ^2).**

7.6 The coefficient of variation

It is sometimes desirable to compare several groups with respect to their relative homogeneity in instances where the groups have very different means. In these circumstances it might be misleading to compare the absolute magnitudes of the standard deviations. It might be more relevant and of greater interest to look at the size of the standard deviation relative to the mean. A measure of relative variability can be obtained by dividing the standard deviation by the mean.

The coefficient of variation (V) is the standard deviation divided by the mean:

$$V = \frac{\sigma}{\bar{x}}$$

COEFFICIENT OF VARIATION

If two groups of data are compared the results could be as follows:

Group A $\bar{x} = 15$, $\sigma = 3$, $V = \dfrac{3}{15} = 0.2$

Group B $\bar{x} = 35$, $\sigma = 5$, $V = \dfrac{5}{35} = 0.14$

The coefficient of variation can be made into a percentage by multiplying by 100. The coefficient of variation for group A then becomes 20% and for group B 14%.

7.7 Conclusions

A measure of dispersion helps to describe a distribution better than an average on its own. It can show whether or not the figures in the distribution are clustered closely together or well spread out.

Dispersion may be as important as an average because in many areas it is changes in the spread of a distribution which are of interest as much as changes in the average. For instance, economists may be as interested in changes in the distribution of incomes as in changes in the average income.

The standard deviation is the most important of the measures of dispersion because of its mathematical properties (especially in sampling theory).

ASSIGNMENTS

1   Obtain a price list for any consumer goods where there are at least ten different prices. Calculate the arithmetic mean, the median, the modal price, the price range and the interquartile range.

   Discuss how far the calculations have helped to summarise the list of prices. Comment on which calculation might be the most useful one to make if you were to consider buying the consumer goods.

2   The following table summarises the performance of two companies on wages, hours worked and days holiday. From this information write a profile of the two companies. On the basis of this evidence, comment on which of these two companies might be the better one to be employed by.

	Company A		Company B	
	Arithmetic mean	Standard deviation	Arithmetic mean	Standard deviation
Weekly wages	80	5	90	15
Weekly overtime earnings	20	10	20	15
Hours worked per week	35	1	40	3
Number of days holiday a year	20	4	15	6

3   A company has the following distribution of overtime earnings. Calculate the arithmetic mean and the standard deviation. Discuss what the value of the standard deviation shows about the weekly overtime earnings of this company.

Weekly overtime earnings (£)	Number of staff
5 and under 10	3
10 and under 15	10
15 and under 20	25
20 and under 25	8
25 and under 30	4

4   The following data shows the monthly expenditure on advertising of the branches of an electrical goods company during a year:

Monthly advertising expenditure (£)	Number of branches
400 and less than 600	4
600 and less than 800	52
800 and less than 1000	110
1000 and less than 1200	70
1200 and less than 1400	64
1400 and less than 1600	47
1600 and less than 1800	43
1800 and less than 2000	10

From this data calculate the median monthly expenditure and explain what it indicates about the branches' advertising expenditure during the year. Also, calculate the interquartile range and explain the purpose of this calculation.

5   The following table shows the results of information collected by a company selling consumer products directly to retailers:

Average number of orders taken per month by individual sales people	Number of sales people
10 and under 20	2
20 and under 30	4
30 and under 40	9
40 and under 50	11
50 and under 60	12
60 and under 70	35
70 and under 80	30
80 and under 90	16
90 and under 100	1

Calculate the range, the standard deviation and the coefficient of variation.

Discuss how far these measures of dispersion help to interpret the information the company has collected.

6   Discuss the purpose of:
(i)   calculating a measure of dispersion;
(ii) identifying differently shaped distributions.

8 Statistical decisions

OBJECTIVES

- To provide an understanding of probability and the normal curve
- To provide an understanding of standard error, tests of significance and confidence limits
- To consider aspects of statistical quality control and statistical estimation

8.1 Estimation

Much of the coverage of the chapters up to this stage has been concerned with the collection of statistical information and maximising the comprehension of the groups of figures collected. The next step is to increase comprehension of the populations from which the figures have been collected.

This is the difference between observing vehicles using a section of road and describing what happens, and using this group as a sample to describe all vehicles including all those not observed.

Statistical estimation is concerned with finding a statistical measure of a population from the corresponding statistic of the sample; for instance, whether the arithmetic mean of a sample is a good estimate of the arithmetic mean of the population. It is an 'estimate' because it is never certain that a sample will be an exact miniature copy of the population itself.

As was seen in Chapter 3, the underlying objective of sampling is to describe the population from which the sample is taken. It was established there that sampling is based on a random selection of items and on the theory of probability. In this chapter it is possible only to provide a very brief discussion of probability and of statistical estimation to give some idea of the basis of statistical induction; that is the process of drawing general conclusions from a study of representative cases.

8.2 Probability

The origins of the concept of probability lie in a simple mathematical theory of games of chance and can be traced back to the 1650s when a French gambler consulted a well-known mathematician, Pascal, presumably to find out how often he might win.

Uncertainty is common to games and to business, and probability can provide analytical tools for measuring and controlling aspects of uncertainty. The use of probability in decision making can be based on an analysis of the past behaviour of sales or production and probabilities assigned to possible outcomes.

This is not the same as the subjective approach used by bookmakers when fixing the odds for a particular horse to win a race. In this approach, initially the odds will be determined by the personal view of the bookmaker, and will be modified as the time for the race approaches according to the subjective views of the punters.

A definition of probability is that the probability of an event is the proportion of times the event happens out of a large number of trials. An 'event' is an occurrence. Obtaining a 'head' when tossing a coin is an event. A failure is also an event, so *not* obtaining a head is an event.

PROBABILITY

1 The probability of a date chosen at random from a calendar being a Monday would be 1 in 7, or 1/7 or 0.143 or 14.3%.
2 The probability of tossing a head when throwing a coin would be 1 in 2 or $\frac{1}{2}$ or 0.5, or 50%.

In 1, if a very large number of dates were looked at, then each day of the week would turn up the same number of times. Therefore Monday would turn up on 1/7 of the occasions and, by definition, this is the probability.

In everyday conversation, words like 'chance', 'likelihood' and 'probability' are used to convey information, but they are not related to the above definition of probability. They are subjective in the same way as the bookmaker's approach; they are estimates based on experience and knowledge of the factors surrounding the situation.

It can be argued that most decisions are made on a qualitative rather than a quantitative basis, even in business. Actions are taken because certain events are very likely to happen, not because their probability is precisely say 97.3%.

As was suggested in Chapter 1, statistics are introduced to support decisions, to provide evidence and to narrow the area of disagreement.

Therefore probabilities can be regarded as relative frequencies: the proportion of time an event takes place. The relative frequency is the probability that the particular event will happen.

1 When it is said in a manufacturing company that the 'probability of an order being complete on time is 0.8', it is meant that on the basis of past experience 80% of all similar orders were met on time.
2 The probability of obtaining a head when a coin is tossed is 50%. This does not mean that, when a coin is tossed ten times, five heads are always obtained. However, if the experiment (or sample or event) is repeated a large number of times then it is likely that 50% heads will be obtained. The greater the number of throws the nearer the approximation is likely to be. In this sense probability is a substitute for certainty.

Empirical definition of probability

The empirical definition of probability is based on experimental data. If a coin is tossed ten times out of which there are seven heads, the empirical probability is $\frac{7}{10}$ or 0.7. As the number of tosses increases, the ratio between heads and tails will become more stable or approach a limit. This limit coincides with the value derived from the 'classical' definition of probability.

Classical definition of probability

$$\text{Probability} = \frac{\text{the number of ways the event can happen}}{\text{the total number of outcomes to the experiment}}$$

Sample space

The sample space is the set of possible outcomes to an experiment.
For example:

Rolling a die: sample space, $S = 1, 2, 3, 4, 5, 6$
Tossing a coin: sample space, $S = H, T$

A coin is tossed twice and the sequence of heads and tails observed: $S = HH, HT, TH, TT$

Measuring probability

The most useful measure of probability involves relating how often an event *will normally occur* to how often it *could* occur – and expressing this relationship as a fraction. If a coin is tossed 1000 times, then normally 500 heads will be obtained. So the probability of obtaining a head is measured as $\frac{500}{1000} = \frac{1}{2}$.

In the case of a pack of cards, the ace of spades will normally be drawn once in 52 draws, so the probability of drawing the ace of spades is $\frac{1}{52}$.

When an event is absolutely certain to happen, the probability of its happening is 1. When an event can never happen the probability of its happening is 0. All probabilities therefore have a value between 0 and 1 (see Figure 8.1).

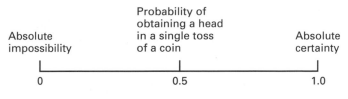

Probability of the event occuring

Figure 8.1 *Measuring probability*

Example 1
A die is rolled; what is the probability of an even score?
The sample space, $S = 1, 2, 3, 4, 5, 6$.
The way the event can happen (the number of even scores): 2, 4, 6 (or three ways).

$P = \frac{3}{6} = \frac{1}{2}$

Example 2
A bag contains five blue balls, three red balls and two black balls. A ball is drawn at random from the bag. Calculate the probability that it will be (a) blue, (b) red or (c) not black.

(a) P (blue) $= \frac{5}{10} = 0.5$
(b) P (rcd) $= \frac{3}{10} = 0.3$
(c) P (black) $= \frac{2}{10} = 0.2$
   P (not black) $= 1 - 0.2 = 0.8$

The probability that the ball will be blue is 0.5 (blue balls constitute half the balls in the bag), that it will be red is 0.3 and that it will not be black is 0.8 (8 is the total number of balls minus the number of black balls), i.e. the other two probabilities added together.

Mutually exclusive events (or complementary events)

Two events are said to be mutually exclusive if the occurrence of one of them excludes the occurrence of the other, i.e. only one can happen. For example, if a coin is tossed and a head appears uppermost, the tail cannot appear uppermost.

If three events are mutually exclusive then the probability of any one event occurring is the sum of the individual probabilities. If A, B and C represent the three events and P the probability of occurrence, then $P(A$ or B or $C) = P(A) + P(B) + P(C)$.

A bag contains four red marbles, two white marbles and four blue marbles. What is the probability of

1 drawing a red marble,
2 not drawing a red marble?

$$1 \quad P = \frac{\text{number of red marbles}}{\text{total number of marbles}} = \frac{4}{10} = \frac{2}{5}$$

$$2 \quad P = \frac{\text{number of blue and white marbles}}{\text{total number of marbles}} = \frac{6}{10} = \frac{2}{5}$$

(The probability of not drawing a red marble is the complement of the probability of drawing red marbles.) With mutually exclusive events, the cases must equal $1 : \frac{2}{5} + \frac{3}{5} = 1$.

Independent events

Two or more events are said to be independent if the occurrence or non-occurrence of one of them in no way affects the occurrence or non-occurrence of the others.

Example 1
If A and B represent the result 'heads' in two successive tosses of a coin, then the events are independent, since the second toss cannot be influenced by what happens before.

When two events, A and B, are independent, the probability that they will both occur is given by:

$$P(A \text{ and } B) = P(A) \times P(B).$$

The answer is the product of the individual probabilities, no matter how many of them there may be.

Example 2
If the probability that one toss of a biased coin will produce heads is $\frac{1}{4}$, the probability that three successive tosses of the same coin will produce heads is $\frac{1}{4} \times \frac{1}{4} \times \frac{1}{4} = \frac{1}{64}$.

Conditional events (or combined events)

Two or more events are said to be conditional when the probability that event B takes place is subject to the proviso that A has taken place. This is usually written $P_A(B)$. $P_B(A)$ would be the other way around, the assumption being that B had taken place first.

If two events A, B are conditional, the probability that one of these events will occur is given by:

$$P(A \text{ or } B) = P(A) + P(B) - P(A \times B).$$

The reason for subtracting $P(A \times B)$ is because A and B are not necessarily mutually exclusive, and therefore both events might occur.

Example
The probability of a firm failing through shortage of capital is 0.6 and the probability of failing through shortage of orders is 0.5.

$$P(A \text{ or } B) = 0.6 + 0.5 - 0.3 = 0.8$$

Tree diagrams

Tree diagrams are used to reduce the need to produce a complete sample space for complicated combinations.

Example 1
What is the probability of drawing two aces from a pack of 52 cards (replacing the first card before drawing the second)?
The probability of the first card drawn being an ace is $\frac{1}{13}$ ($\frac{4}{52} = \frac{1}{13}$). The probability of not drawing an ace is $\frac{12}{13}$ (see Figure 8.2). The replacement of the first card means that the probabilities remain the same for the second draw. If this experiment was repeated a great number of times $\frac{1}{13}$ of the experiments would produce an ace on the first draw. Out of these experiments, $\frac{1}{13}$ would produce a second one. In other words $\frac{1}{13}$ of $\frac{1}{13}$ of the experiments would be expected to produce two aces.

$$P \text{ (drawing two aces)} = \frac{1}{13} \times \frac{1}{13} = \frac{1}{169}$$

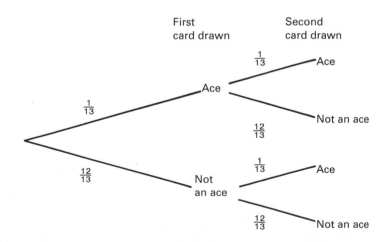

Figure 8.2

> **Example 2**
> A bag contains four red balls and five blue balls. A second bag contains two red balls and five blue balls. If a ball is taken at random from each bag, find the probability that both balls are red.
>
> $$P \text{ (both balls are red)} = \frac{4}{9} \times \frac{2}{7} = \frac{8}{63}$$
>
> (See Figure 8.3.)

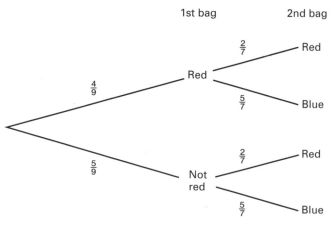

Figure 8.3

8.3 The normal curve

The normal curve is a bell-shaped symmetrical distribution (see Sections 7.1 and 7.4); **its shape will depend on the mean and the standard deviation.** Typical distributions include those that arise in sampling theory (the sampling distribution of the mean), in intelligence tests, in examination results and in biological data such as height and weight. A normal curve may be very tall, or very flat, as shown in Figure 8.4. On both of the curves in Figure 8.4 the arithmetic mean is shown in the centre. In both cases approximately 68% or two-thirds of the distribution will lie within one standard deviation on either side of the arithmetic mean; approximately 95% or 19 out of 20 of the items will lie within two standard deviations on either side of the mean; almost all the items will lie within three standard deviations of the mean (see Section 7.4). Sometimes 2.58 standard deviations on either side of the mean are used to include 99% of the items, while 99.74% of the items are represented by three standard deviations on their side of the mean.

Mathematicians have computed the areas that lies under any part of the curve (a table of these areas is given in Appendix A.2). In the area table, the area between the mean line, the curve, the horizontal axis and a vertical line measured from the mean in units of standard deviations is given as a decimal of the total area enclosed by the curve. For example, if the vertical line lies $1\frac{1}{2}\,\sigma$

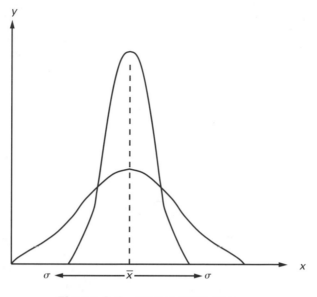

Figure 8.4 Normal distributions

from the mean, the area table shows that the area enclosed is 0.4332 or 43.32% of the total area, or 86.64% for $1\frac{1}{2}\sigma$ on either side of the mean.

The curve which represents these areas is referred to as a 'standard normal distribution'. To apply a standard normal distribution to an actual distribution, the actual distribution is superimposed on the standard normal distribution scale so that the actual distribution mean coincides with the 'zero' (or the mean) and the actual distribution values coincide with the appropriate σ points.

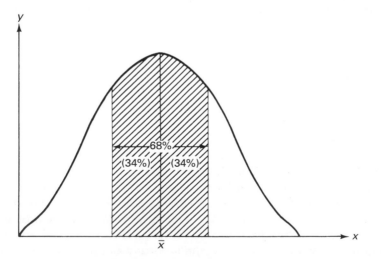

Figure 8.5 Area under the curve

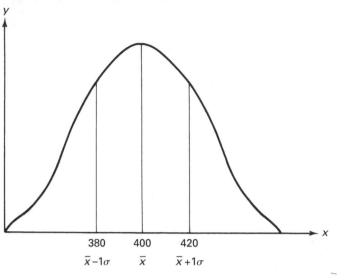

Figure 8.6 A normal distribution

STANDARD NORMAL DISTRIBUTION

A distribution with a mean of 400 cm and a standard deviation of 20 cm will have 400 cm at 0 and the 400 − 20 and 400 + 20 values lying at the −1σ and +1σ points (Figure 8.6).

The standard normal distribution is a way of stating an actual value in terms of its standard deviation from the mean in units of its standard deviation (illustrated in Figure 8.6). This new value can be called a z value. z is the distance any particular point lies from the mean, measured in units of standard deviation:

$$z = \frac{\text{the value} - \text{the mean}}{\text{standard deviation}} = \frac{x - \bar{x}}{\sigma}$$

where x is any particular value (see Figure 8.7).

In the example in Figure 8.6 the z value of 360 cm would be:

$$z = \frac{360 - 400}{20} = \frac{-40}{20} = -2$$

Therefore the 360 cm value lies two standard deviations below the mean of the distribution, and this means that approximately 47.7% of the area under the curve lies between 360 cm and 400 cm (the mean). This is found from the area tables ($z = -2$, area 0.4772 or 47.7%). Therefore if the mean and the standard deviation are known, and what is required is an idea of the percentage of the distribution lying between the mean and a particular value, the z values can be used.

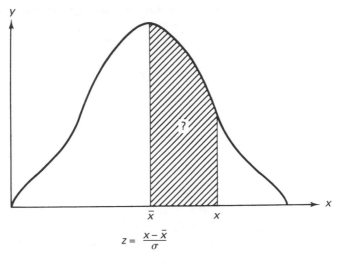

$$z = \frac{x - \bar{x}}{\sigma}$$

Figure 8.7   The z value

---

z VALUES

If a distribution has a mean of 20 and a standard deviation of 5, the area under the curve between the mean and 28 would be given by the z value:

$$z = \frac{28 - 20}{5} = 1.6$$

The area table shows that when $z = 1.6$ the area is 0.4452. Therefore the area lying under the curve is 44.5%, so 44.5% of the distribution will lie between 20 and 28.

8.4   Probability and the normal curve

The fact that 68% of the total area under a normal curve lies within one standard deviation either side of the mean makes it possible to say that two items out of three lie within one standard deviation of the mean. Therefore, if one item was selected at random from a distribution, there would be a 2 in 3 chance of it being one of the items lying within one standard deviation of the mean. This would be correct two times out of three or it would be likely with a '68% level of confidence'.

Similarly, since 95% of the area lies within two standard deviations of the mean, one item selected at random would lie within these limits 19 times out of 20. It would be possible to say the item lay within these limits at a '95% level of confidence'. Nearly all the items would be within three standard deviations of the mean with a '99% level of confidence'.

Therefore, **if a sample is selected at random from a population, there is a good chance or probability that it will represent the population from which it is drawn.** Describing a population from observing a small sample is not a matter of guesswork provided:

- the sample is taken at random,
- it is understood that there is always some degree of error in sampling.

A sample will not be an exact replica of the population, because chance plays an important part in random selection. It is likely that the sample mean and standard deviation will differ a little from the population mean and deviation. The bigger the sample the less chance there is of selecting a group of items that is unrepresentative of the population.

In general it can be stated that:

- **the best estimate of the population mean is the sample mean;**
- **the best estimate of the population standard deviation is the sample standard deviation.**

8.5   Standard error

The fact that there is always some degree of error in sampling means that it is useful to have a measure which provides an indication of the extent to which sample means deviate from population means.

If a large number of samples are taken from a population that is normally distributed, most of the sample means will be the same or very similar. Occasionally a sample will by chance contain an undue number of high or low values, so that its mean will be above or below other means. If all these means are graphed they will look like a normal curve, with most of the means in the centre but some on either side. **This 'sampling distribution of the mean' will be normally distributed and the average of these means will be equal to (or 'the best estimate of') the true population mean.**

The standard deviation of the sampling distribution of the mean is called the standard error. It is found by the formula:

$$\text{standard error (S.E.)} = \frac{\text{standard deviation of the sample}}{\sqrt{\text{sample size}}}$$

$$= \frac{\sigma}{\sqrt{n}}$$

where n is the sample size.

(In this formula σ may be replaced by S because this is the estimate of the population standard deviation.)

Therefore:

- since 95% of the items in a normal distribution lie within two standard deviations of the mean of the distribution,
- since the distribution of the means is normal with a mean equal to the population mean,
- 95% of the means of all samples must lie within two standard errors of the true mean of the population,
- and therefore, if a single sample is taken, 19 times out of 20 the sample mean will lie within two standard errors of the true mean of the population. Or, in other words, 19 times out of 20 the true mean of the population cannot lie more than two standard errors from the mean of the sample.

Notice that the accuracy of the estimate is independent of population size. A population of 2 million does not require a bigger sample than a population of 20 000. Accuracy depends on the sample size and the variability of the characteristics measures (see Section 3.9).

SAMPLE SIZE

A sample of 100 with a standard deviation of 5 will have a standard error of:

$$\text{S.E.} \frac{\sigma}{\sqrt{n}} \frac{5}{\sqrt{100}} = \frac{5}{10} = 0.5$$

The accuracy of the sample can be increased by expanding the size of the sample. A sample of 200 will have a standard error of:

$$\text{S.E.} = \frac{5}{\sqrt{200}} = \frac{5}{14.2} = 0.35$$

To halve the standard error in the original sample (0.5), the sample size has to be increased by four times:

$$\text{S.E.} = \frac{5}{\sqrt{400}} = \frac{5}{20} = 0.25$$

The variability of any given population may change over time. When the electorate of the UK was felt to be predictable in its voting patterns (in the 1950s and 1960s) a sample of 2000 electors was thought to provide a reasonably accurate prediction of election results. In more recent years it has been felt that the variability and volatility in the electorate has increased and therefore larger samples are required to provide a reasonable prediction of election results. *Notice that this is not because of an increase in the size of the population but because a larger sample is required to represent this population.*

The theory behind standard error and sampling distributions is part of a mathematical theory called the central limit theorem (see Section 3.9). Detailed discussion of this theorem lies outside the scope of this book. However, as a result of this theorem it can be accepted that the sampling distribution is normal even if the population frequency distribution is not normal, provided that the sample size is sufficiently large (greater than 30). This means that it is possible to take a sample from any population and apply methods of estimation.

8.6   Tests of significance

Sometimes a fact or theory is believed to be true, but when a random sample is taken the results do not wholly support the fact or theory.

The difference between the belief or hypothesis and the sample result may be because:

- the original theory or hypothesis was wrong,
- the sample was one-sided.

Significance tests are aimed at revealing whether or not the difference between a hypothesis and a sample result could reasonably be ascribed to chance factors operating at the time the sample was selected. If the difference cannot be explained as being due solely to chance the difference between the theory and the sample result is said to be statistically significant.

To test a hypothesis it is possible:

- to collect information on the whole population and then accept or reject the hypothesis with complete certainty;
- to take a random sample, if the population is too large for a full survey, and test the null hypothesis.

The 'null hypothesis' is the assumption that there is no difference between the hypothesis and the sample result.

SIGNIFICANCE TEST

If the theory is that a group of employees receive average overtime of £130 a month, then the null hypothesis (Ho) is that the population mean (μ: pronounced 'mew') is £130. The opposite theory (Hi) is that it is not £130:

Ho: μ = £130 (the population mean is £130)
Hi: $\mu \neq$ £130 (the population mean is not £130)

A random sample is taken and the results show:

$n = 100$ $\bar{x} = £123$ $\sigma = £30$

If the population mean is £130, then 95% of the means of all samples will fall within two standard errors of this figure (as was seen in the sampling distributions of the means). If the sample mean, £123, is not within two standard errors of £130, then the population mean is probably not £130 unless the sample mean is the one in twenty, that is one-sided.

The standard error of the sample results is:

$$\text{S.E.} = \frac{\sigma}{\sqrt{n}} = \frac{30}{\sqrt{100}} = \frac{30}{10} = 3$$

The critical values are the limits in which the sample mean will fall 95% of the time if the population mean is correct.

The values are found by taking the population mean, plus and minus two standard errors:

$$\text{critical values} = £130 \pm 2 \times 3$$
$$= £130 \pm 6$$
$$= £124 \text{ and } £136$$

The sample mean is £123. This does not lie within the critical values; therefore Ho (the null hypothesis) is rejected at the 5% level of significance. The difference between the assumed population mean (£130) and the sample mean (£123) can be said to be significant.

From this test it can be stated that there is evidence to suggest that the population mean is not £130 a month.

If the sample mean had fallen within the critical values the Ho would be accepted at the 5% level and the difference between μ and $\bar{x}$ would not be significant. This would not mean that the hypothesis was 'proved'. It would mean that there was evidence to suggest that the population mean could be £130 a month.

It is possible to expand the critical values by taking the 1% level of significance where it is suggested that between 99% and 100% of the items (sample means) will lie within three standard errors of the population mean:

$$\text{critical values} = £130 \pm 3 \times 3$$
$$= £130 \pm 9$$
$$= £121 \text{ and } £139$$

At this level of significance the null hypothesis (Ho) is accepted. Therefore there *is* evidence to suggest that the population mean could be £130 a month, because the sample mean (£123) falls within the critical values.

However, by widening the critical values there is an increased probability that the null hypothesis is being accepted when it is in fact wrong.

There are two types of error which can be made in testing significance which are called type I and type II:

- **a type I error is rejecting the null hypothesis (Ho) when it is in fact true,**
- **a type II error is accepting the null hypothesis when it is in fact false.**

The level of significance can be said to equal the probability of making a type I error. If the critical values are expanded there is an increased probability of making a type II error.

The level of significance chosen and the interpretation of the results of the test are a matter of judgement. It is important that the level of significance (1%, 5% or another level) is decided before data is collected or calculations are made, to avoid the possibility of manipulating the significance level to produce a particular result. If more evidence is required, further random samples can be taken.

8.7　Confidence limits

Instead of testing specific values of the population mean as a significance test, **an alternative test is to construct an interval that will, at specified levels of probability, include the population mean.**

CONFIDENCE LIMITS

A random sample taken from the hourly rates of pay of a group of self-employed people provides the following results:

$$n = 100 \quad \bar{x} = £200 \quad \sigma = £20$$

These results can be used to set 95% confidence limits on the unknown population mean μ. The standard error of the sample results is:

$$\text{S.E.} = \frac{20}{\sqrt{100}} = \frac{20}{10} = 2.$$

$$
\begin{aligned}
95\% \text{ confidence limits} &= \bar{x} \pm 2 \times \text{S.E.} \\
&= £200 \pm 2 \times 2 \\
&= £200 \pm 4 \\
&= £196 \text{ and } £204
\end{aligned}
$$

It can be said (approximately) that there is a 95% chance that this range will include the unknown population mean (μ).

Narrower limits can be achieved by increasing the size of the sample. If the sample size was 400 and the results were the same, then:

$$\text{S.E.} = \frac{20}{\sqrt{400}} = \frac{20}{20} = 1$$

$$
\begin{aligned}
95\% \text{ confidence limits} &= £200 \pm 2 \times 1 \\
&= £198 \text{ and } £202
\end{aligned}
$$

Setting 99% confidence limits increases the likelihood of including the population mean in the range, but it also increases the size of the interval.

With $n = 100$:

$$
\begin{aligned}
99\% \text{ confidence limits} &= £200 \pm 3 \times 2 \\
&= £194 \text{ and } £206
\end{aligned}
$$

With $n = 400$:

$$
\begin{aligned}
99\% \text{ confidence limits} &= £200 \pm 3 \times 1 \\
&= £197 \text{ and } £203
\end{aligned}
$$

8.8 Statistical quality control

One application of sampling methods and of probability is statistical quality control. This involves sampling the output of a manufacturing process to ensure that the quality of the commodities being produced conforms to specified standards.

Mass production techniques involve making sure that each item of output has standard dimensions within certain limits of an ideal standard. Products can be checked when they come from the machine, but by that time the damage may have been done and a whole batch may be lost. It is more economical to take a series of samples during production in order to try to discover any fault at the earliest possible moment.

With any commodity there may be variations in a number of dimensions of the commodity, such as the length. Frequent samples, usually taken automatically, will check on this. If a frequency distribution of the mean length of products is taken, the curve is likely to be bell-shaped, or similar to the normal curve. Most components will be the right size but a few will be too large or too small.

QUALITY CONTROL

If the ideal length of a component is 5 cm and there is a tolerance of 0.004 cm, then components of 4.996 cm and of 5.004 cm will be acceptable.
The frequency curve of a sample may appear as in Figure 8.8.

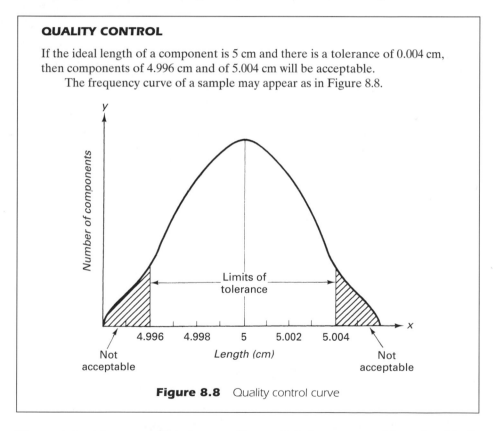

Figure 8.8 Quality control curve

The samples taken are plotted on quality control charts to provide a visual indication of quality variations (Figure 8.9). In Figure 8.9 the action or control limits are limits of standard error and therefore are confidence limits. One important

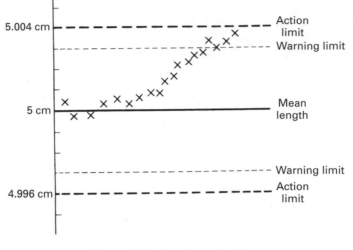

Figure 8.9 Quality control chart

advantage of the control chart is that output from a production process often deteriorates progressively and this is made apparent in a chart by the trend of the plotted points. In the example in Figure 8.9 the points are moving out towards the upper warning limit and therefore the production line should be investigated.

More detailed and complicated techniques can be applied to the control of quality and important aspects of quality control in practice involve engineering problems. It is introduced in this chapter as an example of sampling.

8.9 Conclusion

It can be argued that decisions are often made on emotional and irrational grounds rather than on facts. Perhaps this depends on the type of decisions being made. For instance, if couples looked at the facts (costs, sleepless nights, etc.) before they decided to have children, the population might decline rapidly.

The decision to buy a car or a house has to be based on some facts. For instance the amount of money available through saving and borrowing must be a limiting factor. The size of the family and other factors may also influence the choice of car or house. Once these factors have been taken into account, the choice of a particular car or house may depend on personal preferences or chance factors and therefore not on clearly rational grounds.

Justification for such decisions will tend to be expressed in rational terms although sometimes people will admit that they bought a car because they liked the colour or shape, or a house because they 'fell in love with it'.

Business decisions are perhaps similar to this. The facts provide the limits within which preferences and 'hunches' have to work; and the facts provide the justification for the decision when it is being explained to the bank manager, the 'boss' or the shareholders.

Statistical decisions are concerned with the reliance which can be placed on facts, the confidence with which figures can be used and the interpretation of facts and figures.

ASSIGNMENTS

1   Discuss the problems of decision making. How are decisions made in business? How far can statistics help in business decision-making?
2   What is the meaning of standard error? Why is the concept important in sampling?
3   It is thought that the cost of transport to and from work of a group of employees is £100 a month. A sample is taken and the results show that, of the 100 employees surveyed, their average monthly transport costs are £80 with a standard deviation of £10. Is there any evidence from the sample results that the employees' transport costs could be £100 a month?
4   Discuss the importance in statistics of:
    (i)   probability,
    (ii) the normal curve.
5   Comment on the use of significance tests and confidence limits. How far do they help in making decisions?
6   Find examples of the use of quality control. Consider the methods used in these cases.

9 Comparing statistics: index numbers and vital statistics

OBJECTIVES

- To discuss the use and usefulness of index numbers
- To provide an understanding of the calculation of a weighted index number and a chain-based index
- To consider some of the problems of index number construction

9.1 Index numbers

An index number is a measure designed to show average changes in the price, quantity, or value of a group of items over a period of time.

The aim is to provide a device to simplify comparison over time, by replacing complicated figures by simple ones calculated on a percentage basis. Most index numbers are weighted averages and they have similar advantages and disadvantages to averages. For this reason many types of index have been developed, some of which are more suitable than others for particular purposes.

INDEX NUMBERS

The average price of houses can be compared between two years:

	Average house prices (£)	*Index number (price relative to year 1)*
Year 1:	50 000	$\dfrac{50\,000}{50\,000} \times 100 = 100$
Year 5:	70 000	$\dfrac{70\,000}{50\,000} \times 100 = 114$

In this index, year 1 is the base year (the year on which the price changes are based) and is written: year 1 = 100. The index for year 5 (114) indicates that average house prices have risen by 14% between year 1 and year 5.

There are three general types of indexes (or indices):

- **price indexes,** which measure changes in prices;
- **quantity indexes,** which measure production and output changes;
- **value indexes,** which measure changes in the value of various commodities and activities.

Where one item is involved in comparisons between different periods of time, the price or value index is found by:

$$\text{price index} = \frac{p_1}{p_0} \times 100$$

where p_0 is the price in the base year and p_1 the price in the year to be compared. Where more than one item is involved, the calculation of an index is more complicated. In calculating an index to show changes in the cost of living, for example:

- some prices will rise, others will fall;
- prices are for different units (weights and quantities of goods);
- different products are of varying importance in the cost of living;
- households spend their money in different ways.

PRICE INDEX

An extreme case would be to measure changes in the cost of living by changes in the prices of two commodities, say bread and matches. It has to be assumed that there is an 'average' loaf of bread of a particular weight and size and an 'average' box of matches containing a certain number of matches.

If the price of bread has risen from 80p for an average loaf in year 1 to £1.60 for an average loaf in year 3 this is a percentage increase (year 1 = 100) of 50%.

If the price of matches has increased from 20p for an average box in year 1 to 24p for an average box in year 3 this is a percentage increase (year 1 = 100) of 20%.

Therefore the average percentage increase in the cost of living could be calculated as:

$$\frac{50 + 20}{2} \% = 35\%$$

Or it could be said that the index had risen from 100 in year 1 to 135 in year 3.

However, there are a number of problems with both these statements.

1 They are based on assumptions such as the choice of only two commodities, and the 'average' size.
2 Household A may consume large amounts of bread but never buys matches so it has had an actual increase in the cost of living of 50%, and the 35% figure does not represent it at all.
3 Household B may consume small amounts of bread and buy several boxes

of matches a week. No allowance has been made for the various amounts bought of each. Even if this household buys three boxes of matches to one loaf of bread, it is still spending more on bread.

This illustrates the problem of trying to provide an average figure to represent changes involving large numbers of people and households or companies and goods. Despite these problems a very large number of indexes are produced, for prices, wages, production, sales, transport costs, share prices, imports and exports and so on. These indexes can be useful for several reasons.

1   They can provide background information against which business people can compare their performance and which provide material for government decisions.
2   Pay and pensions are 'indexed' by being automatically tied to the Index of Retail Prices, so that the purchasing power of money income and pension is preserved at times of rising prices.
3   Index numbers can be used to make projections to ascertain likely future changes in prices, output and so on. These comparisons and projections facilitate decision making at company, industry and government level.

9.2   Weighted index numbers

In an attempt to overcome the problem of the difference in importance of items in an index and different units of measurement, **a weighted index number makes the figures directly comparable.**

The weights reflect the relative importance of an item. If food is given a weight of 300 points and housing a weight of 150 it means that a change in the price of food is twice as important as a change in the price of housing.

Given a system of weighting, an index number can be calculated by the price relative method (Table 9.1). This is known also as the Weighted Average of Price Relatives Index, or Average of Weighted Price Relatives Index.

Table 9.1   Price relative method.

Item	(1) Base price (£) (year 1)	(2) Price (£) (year 3)	(3) Percentage increase (or price relative)	(4) Weight	(5) Product (3) × (4)
Food	25	30	20	300	6 000
Housing	20	22	10	150	1 500
Transport	5	10	100	100	10 000
Services	10	12	20	50	1 000
				600	18 500

In Table 9.1, the increase in price in year 3 (column 2) over year 1 (column 1) is calculated as a percentage (column 3). This is found by taking the difference in prices over the year 1 price and multiplying by 100 to arrive at the percentage increase (or decrease).

For example, for food and housing the percentage increases are:

$$\frac{5}{25} \times 100 = 20\%$$

and

$$\frac{2}{20} \times 100 = 10\%$$

The percentage increase in price is then multiplied by the weight (column 4) to arrive at the product (column 5). For example:

$$20 \times 300 = 6000$$

and

$$10 \times 150 = 1500$$

All the products are added together (18 500), and the result is divided by the sum of the weights (600):

$$\text{The weighted average} = \frac{18\,500}{600} = 30.83\%$$

Therefore, on average, prices in year 3 have risen by 30.83% over prices in year 1. If the index was assumed to be 100 in year 1 it would be 130.83 in year 3.

The formula for calculating the index number is:

$$\frac{\sum\left(\dfrac{p_1}{p_0} \times 100 \times w\right)}{\sum w}$$

where p_0 is the base price, p_1 the new price and w the weighting, and the calculation in the bracket is made for each item.

In the example

$$\sum\left(\frac{p_1}{p_0} \times 100 \times w\right) = \left(\frac{30}{25} \times 100 \times 300\right) + \left(\frac{22}{20} \times 100 \times 150\right)$$

$$+ \left(\frac{10}{5} \times 100 \times 100\right) + \left(\frac{12}{10} \times 100 \times 50\right)$$

$$= 36\,000 + 16\,500 + 20\,000 + 6000$$

$$= 78\,500$$

$$\Sigma\left(\frac{p_1}{p_0} \times 100 \times w\right) = \frac{78\ 500}{600}$$

$$= 130.83$$

If the weighting system is changed (Table 9.2) a new index can be calculated.

Table 9.2

Item	(1) Percentage increase from year 1 to year 3	(2) Weight	(3) Product (1) × (2)
Food	20	220	4 400
Housing	10	150	1 500
Transport	100	200	20 000
Services	20	30	600
		600	26 500

With the new weightings the calculations give:

$$\text{weighted average} = \frac{26\ 500}{600} = 44.17\%$$

$$\text{index number} = \mathbf{144.17}$$

The new weights produce a higher index number because transport has become a more important item and has had the fastest rise in prices.

SOME COMMONLY USED TYPES OF INDEX NUMBER

The Laspeyres index uses base-year quantities and weights. This indicates how much the cost of buying base-year quantities at current-year prices is compared with base-year costs. Different years can be compared with each other.

formula: $\dfrac{\Sigma p_1 q_0}{\Sigma p_0 q_0} \times 100$

where p_0 and p_1 are the base-year and current-year prices, and q_0 is the base-year quantity.

The Paasche index uses the current-year quantities and weights. This indicates how much current-year costs are related to the cost of buying current-year quantities at base-year prices. This index requires actual quantities to be ascertained for each year of the series. The different years can be compared only with the base year and not with each other.

formula: $\dfrac{\Sigma p_1 q_1}{\Sigma p_0 q_1} \times 100$

Fisher's 'ideal index' involves the averaging of the full Laspeyres and Paasche indexes, using the geometric mean. This is a complicated process and is seldom used because it is not practicable. Fisher's index was developed because of the imperfections of the Laspeyres and Paasche systems of index numbers which have led to attempts to find compromise solutions. These have usually involved the averaging of some component of the base and current indexes.

9.3 Chain-based index numbers

In the chain-based index number system each period in the series uses the previous period as the base. Therefore the figures for one year provide the base for the next year. This is a useful system where information on the immediate past is more important than information relating to the more distant past, because it indicates the extent of change from year to year. Also this system emphasises the rate of change. In Table 9.3 the rate of change was much less from year 3 to year 4 than from year 2 to year 3.

Table 9.3 Conversion from a fixed to a chain base.

Year	Fixed index	Chain-based index
1	100	100
2	110	$\frac{110}{100} \times 100 = 110$
3	125	$\frac{125}{110} \times 100 = 113.6$
4	130	$\frac{130}{113.6} \times 100 = 114.4$

9.4 Problems in index-number construction

Choice of items

The choice of items to be included in an index can present problems. If every item is included, the construction of the index may become over-complicated. Some items may have a small influence on the index. On the other hand if items are omitted then some people's interests will be ignored. For example, the Index of Retail Prices in the UK tends to leave out very exotic and luxurious goods, so people who spend their money on hang-gliding or pearl necklaces may feel unrepresented by the index.

Choice of weights

The choice of weights is a considerable problem because the weights are an attempt to reflect the importance of items. For example, the Index of Retail Prices is concerned with price changes for a 'typical household' and like an 'average person' this does not exist. Therefore the weights may not reflect exactly the importance of items for any particular household, although they may be close to a large number.

Choice of base year

The choice of the base year can present problems because of the 'bias' that can arise from the choice. If a 'depressed' year is chosen, the years that follow may appear particularly good; if a 'prosperous' year is chosen, the years that follow may appear poor. What is needed is a 'normal' year, which may be hard to find. At least boom and depression years can be avoided.

Example
Unemployment may appear high if a year of low unemployment is chosen as the base year; and it may appear to have fallen dramatically in subsequent years if a depression year is chosen.

Also, the base year should not be too far in the past or it will be out of date and will appear unrealistic. Therefore base years are changed from time to time. These lead to a further problem.

Comparing indexes

Comparing indexes based on different years is difficult. A change of base year brings an index up to date, but breaks the continuity of a series. This can be overcome by providing comparative figures using the old base for a time.

However, comparisons of index numbers based on different years is often very difficult because of possible changes made in the construction of the numbers, changes in the weights and changes in 'expectations'. Comparisons can suffer from the problems of historical perspective.

Example
It is often stated that a particular period is more difficult for 'first-time house-buyers', because prices have risen so rapidly. Comparisons between house prices and the income of first-time house-buyers may show that the ratio has remained fairly stable. Assuming this is the case it would not be the whole story, because the number of people who want to buy a house and who expect to be able to buy a house may have increased and their expectation of the standard of house they should be able to buy may have risen.

9.5 The Index of Retail Prices

The Index of Retail Prices is probably the best-known index number because it is of interest to everybody as the 'cost-of-living' index. In the UK the index **'measures the change from month to month in the average level of prices of the commodities and services purchased by nearly nine-tenths of the households in the United Kingdom'** (Department of Employment).

The prices of some 350 goods and services are regularly collected and approximately 150 000 separate price quotations are used each month in compiling the index. The expenditure pattern on which the index is based is revised each year using information from the Family Expenditure Survey (FES). Up to 1974 the weighting pattern was established on the base dates (June 1947, January 1952, January 1956, January 1962), but in 1974 it was decided that the weighting pattern should be revised annually in January on the basis of the information obtained from the FES.

The Retail Prices Index measures the change in the cost of a representative basket of goods and services. The composition of this basket (the relative weights attached to the various goods and services it contains) is based on the FES.

The FES is an annual survey based on a three-stage sample, with a stratified sample used at each stage (see Section 3.14). Each household has an equal chance of selection and the weights are based on the 'average' household's basket of goods. Items are given weights out of a total of 1000. The relative importance of an item depends on how necessary it is, changes in the price of it and other commodities and changes in income. Housing and transport have both risen in importance steadily through the years. Food has fluctuated as a result of changing prices and changes in income. As the income of households rises, the same amount is spent on food and more on other items. Services have declined in 'importance' largely because of changes in definition. New categories have been introduced into the weighting system, such as transport.

As a cost-of-living index, the Index of Retail Prices is widely quoted, although it has to be used with some caution in this context (see Sections 9.1 and 9.4). The figures provide no more than a general and average indication. For particular households inflation may have been greater or less than this depending on their patterns of expenditure. Although the Index of Retail Prices cannot provide a completely accurate and comprehensive account of inflation, it does provide the best available indicator of changes in the cost of living and the level of inflation in the UK.

9.6 Vital statistics

Vital statistics are figures relating to births, marriages and deaths. Demography is the statistical study of life in human communities, and the population in these communities, which depends on birth, deaths, immigration and emigration.

These four factors depend on a range of other factors: the birth rate, for example, depends on the sex ratio, the proportion of women of child-bearing age (normally considered to be between 15 and 45), the expected size of the family, the rate of illegitimate births and other social and economic factors.

Crude birth and death rates

Crude birth and death rates are the total (live) births or deaths during 12 months per 1000 of the population.

$$\text{Crude birth rate} = \frac{\text{total live births} \times 1000}{\text{total population}}$$

$$\text{Crude death rate} = \frac{\text{total deaths} \times 1000}{\text{total population}}$$

CRUDE BIRTH AND DEATH RATES

Number of live births = 750 000
Number of deaths = 560 000
Total population = 50 million

$$\text{Crude birth rate} = \frac{750\,000 \times 1000}{50\,000\,000}$$

$= \mathbf{15}$ births per thousand of the population

$$\text{Crude death rate} = \frac{560\,000 \times 1000}{50\,000\,000}$$

$= \mathbf{11.2}$ deaths per thousand of the population

Standardised birth and death rates

Standardised birth and death rates involve the standardisation of statistics so that like can be compared with like. Standardisation can be achieved by, for example, the use of weights. This may be needed if the crude rates for two populations cannot be compared properly because the populations have very different characteristics.

For example, two towns may have very different age profiles. Town A may be full of young people and young families, while town B is full of retired people. In order to compare, for example, the death rates of the two towns on a similar basis the death profiles of the towns can be weighted by the death profile of a standard population, such as that for England and Wales (see Table 9.4).

$$\text{Standardised death rate} = \frac{\text{total of weighted deaths} \times 1000}{\text{total population}}$$

Although from the crude death rate it may appear that town A is much healthier than town B, when the standardised death rates are compared it is clear that

Table 9.4 Death rates for two towns

(1) Age group	(2) Population (thousands)	(3) No. of deaths	(4) Death rate per 1000	(5) Population (thousands)	(6) No. of deaths	(7) Death rate per 1000	(8) Standard population	(9) A (4) × (8)	(10) B (7) × (8)
	Town A			*Town B*				*No. of expected deaths in standardised population if subject to mortality of towns A and B*	
0–9	20	60	3	5	30	6	10	30	60
10–19	20	20	1	10	10	1	5	5	5
20–29	40	80	2	10	20	2	15	30	30
30–39	26	104	4	12	84	7	20	80	140
40–49	12	120	10	23	230	10	18	180	180
50–59	8	240	30	20	200	10	20	600	200
60–69	5	120	24	15	630	42	10	240	420
70+	1	90	90	5	450	90	2	180	180
	100	830		100	1670		100	1345	1215

town B could in fact be said to be healthier than town A. The difference in the crude rates appears to be due to the more youthful population of town A.

$$\text{Town A: Crude death rate} = \frac{830 \times 1000}{100\ 000} = 8.3 \text{ per thousand}$$

$$\text{Town B: Crude death rate} = \frac{1670 \times 1000}{100\ 000} = 16.70 \text{ per thousand}$$

$$\text{Town A: Standardised death rate} = \frac{1345 \times 1000}{100\ 000} = 13.45 \text{ per thousand}$$

$$\text{Town B: Standardised death rate} = \frac{1215 \times 1000}{100\ 000} = 12.12 \text{ per thousand}$$

ASSIGNMENTS

1 Compare the latest available figure for the Index of Retail Prices with the index five years before. How good is the index as an indication of changes in the cost of living?
2 Write a report on the UK Indexes of Distribution.
3 A firm uses four raw materials in its production. Calculate a weighted index number from the information in the table below:

Raw materials	Base price (year 1) (£)	New price (year 3) (£)	Weight
A	35	40	253
B	28	35	124
C	8	20	135
D	12	16	54

Write a report on the results of the calculations.
4 Discuss the problems involved in constructing an index number.
5 Wages, pensions, prices and savings can be linked to an index. Consider the advantages and problems involved in index-linking.
6 The data in the table below relates to two towns A and B in the same year.

	Town A		Town B		
Age group	Population	No. of deaths	Population	No. of deaths	Standard population
0–17	6000	30	4000	20	30%
18–44	6000	20	5000	20	30%
45–64	4000	45	6000	30	25%
65+	4000	120	5000	140	15%

Calculate:

(i)   the death rate for each age group in the towns A and B;

(ii) the crude death rate for towns A and B;

(iii) the standardised death rate for towns A and B.

Comparing statistics: correlation

10.1 What is correlation?

Correlation is concerned with whether or not there is any association between two variables. If two variables are related to any extent, then changes in the value of one are associated with changes in the value of the other.

ASSOCIATION

An increase in the sales of a company may show a strong association with increases in the money spent on the advertising of its products.

Connections between two variables are an everyday occurrence (for instance: the income and expenditure of a household, the costs and sales of a company, the miles driven in a car and the petrol bought for it).

It is useful to know:

- whether any association exists,
- the strength of the association,
- the direction of the relationship,
- the proportion of the variability in one variable that can be accounted for by its relationship with the other variable.

It is useful to try to measure these factors because:

- knowledge of the relationships enables plans and predictions to be made,
- past evidence can be used to make decisions,
- plans and predictions based on evidence rather than guesswork provide great control over events.

PREDICTION

If it is known how many kilometres or miles a commercial vehicle travels using a litre of petrol, it is possible to predict how much petrol will be required for a particular journey.

Also, if it is assumed that the kilometres travelled by a vehicle and the petrol consumed are strongly related, then a sudden change in petrol consumption may provide an early indicator of a problem in the engine.

If it is assumed that a particular vehicle always consumes 5 litres of petrol to cover 50 kilometres, then the association would appear on a graph as in Figure 10.1.

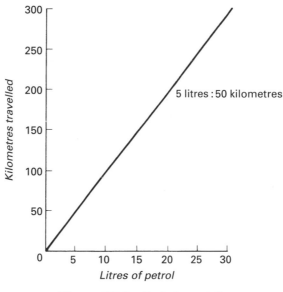

Figure 10.1 Perfect correlation

Figure 10.1 is an example of perfect correlation or association. The fact that the 'curve' is a straight line makes it possible to read off one variable against the other (see Section 5.10). For example, if a journey was to be undertaken of 220 kilometres, it would be clear that the vehicle would consume 22 litres of petrol. If there were 12 litres of petrol in the vehicle it would be known that it could travel exactly 120 kilometres on this amount of petrol.

In fact many relationships between two variables are linear (that is, based on a straight line). There are many non-linear relationships as well and the calculation of the correlation coefficient in these cases is complicated. In considering correlation, trends and time series, the main concern here is with linear relationships.

Few linear relationships are perfect because other factors are involved. For example, if a company spends more on advertising, this does not guarantee a rise in sales and certainly any rise in sales that does take place will not be by a predetermined amount. Competition from other companies, the economic situation, the success with which the advertising is carried out, may all influence the connection between these two variables.

There are a number of methods of indicating correlation between two variables. These include:

- scatter diagrams,
- correlation tables,
- the product moment coefficient of correlation,
- the coefficient of rank correlation,
- regression.

10.2   Scatter diagrams

Scatter diagrams (or scatter graphs) provide a useful means of deciding whether or not there is association between variables.

The construction of a scatter diagram is by drawing a graph so that the scale for one variable (the independent variable, if this can be determined) lies along the horizontal axis, and the other variable (the dependent variable) on the vertical axis. Each pair of figures is then plotted as a single point on the graph.

The basic purpose is to see whether there is any pattern among the points. If there is, a 'line of best fit' can be drawn with the same number of points or co-ordinates on each side of the line. This line is the one judged to be the best line to fit the pattern of points.

In Figure 10.2 the correlation indicated by the line of best fit is approximately linear. Also it indicates positive correlation, that is as advertising expenditure is increased sales revenue tends to rise.

This is not 'proved' by this graph, nor is any causal relationship clear from it. All that can be said with certainty is that there appears to be some association between the two variables. The scatter diagram provides an indication which may be the first piece of evidence in an investigation. A scatter diagram can indicate the type of correlation that exists (see Figures 10.3, 10.4, 10.5, 10.6 and 10.7):

- Figure 10.3 indicates positive correlation: as the variable x increases y will increase.
- Figure 10.4 indicates negative correlation: as x increases y will decrease.
- Figure 10.5 indicates perfect positive correlation between the two variables: they both increase in the same proportions.
- Figure 10.6 indicates perfect negative correlation between the two variables: as one rises the other falls in exact proportion.

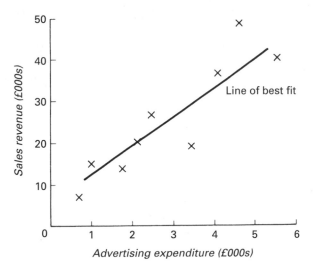

Figure 10.2 Scatter diagram: advertising expenditure and sales revenue of a company

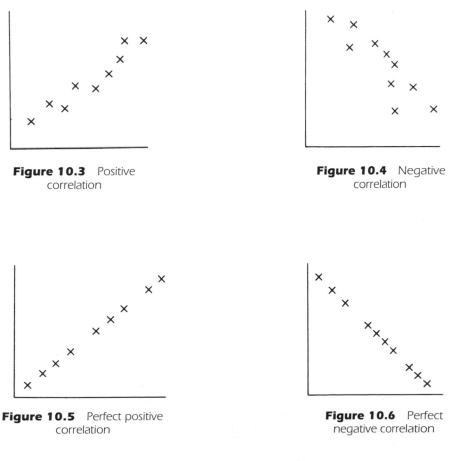

Figure 10.3 Positive correlation

Figure 10.4 Negative correlation

Figure 10.5 Perfect positive correlation

Figure 10.6 Perfect negative correlation

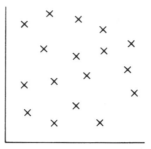

Figure 10.7   No correlation

- Figure 10.7 indicates that there is no correlation between the two variables. There are a number of lines of best fit which could be drawn with equal validity.

Once the line of best fit has been drawn a scatter diagram can be used for estimating by reading off the line the value of one variable against the other. However, this is not likely to be very reliable unless the two variables are very closely related.

Scatter diagrams are limited because:

- there is uncertainty about the correct position of the line of best fit,
- although there is an indication of the type of correlation, there is not any indication of the extent of the correlation.

10.3   Correlation tables

Correlation tables perform a similar function to scatter diagrams. They provide an approximate indication of correlation.

Table 10.1 shows a comparison between the age of a company's employees and their length of service. As would be expected, the older the employees the longer their service has been with the company. In other words there is an indication of positive correlation. The table shows, for example, that:

- 110 out of 200 employees are over 46 years old (55%) and therefore that the company has a fairly old labour force,
- only 20% of the labour force are below the age of 36,
- nobody over the age of 36 has been newly employed by the company for at least five years.

Although this type of information can be seen from the table, the information it provides on correlation is limited.

Table 10.1 A correlation table.

Age of employees (to the nearest year)	Length of service (to the nearest year)						
	Under 5	5–9	10–14	15–19	20–24	25–29	30 and above
16–25	14	7					
26–35	1	18					
36–45		2	28	18	1		
46–55			10	22	36		
56–65		1			24	12	6

10.4 The product moment coefficient of correlation

The 'product moment' coefficient is a measure of linear correlation (devised by Karl Pearson and sometimes referred to as Pearson's coefficient), which attempts to show the closeness of the relationship between two variables. This is not only an attempt to show whether correlation exists (which is also the aim of scatter diagrams and correlation tables), but also an attempt to show how closely it exists.

The coefficient is based on a formula which divides the mean product of the deviations from the mean (shown by $\Sigma(xy)$) by the product of the standard deviations (shown by $\sigma_x\sigma_y$):

$$r = \frac{\Sigma(xy)}{n\sigma_x\sigma_y}$$

where r is the product moment coefficient
x is one variable (which is sometimes referred to as the subject)
y is the other variable (which can be referred to as the relative)
n is the number of items (subject or relative, not both)
σ_x is the standard deviation of variable x
σ_y is the standard deviation of variable y

This formula can be rewritten on the basis of the basic formula for the standard deviation:

$$r = \frac{\Sigma(xy)}{\sqrt{\Sigma x^2 \Sigma y^2}}$$

where $\Sigma(xy)$ represents the sum of the products of the deviations from the mean

     x is the deviation from the mean of one variable

     y is the deviation from the mean of the other variable

Table 10.2   Product moment coefficient.

Year	Investment (£000s)	Profit (£000s)	x (deviation from $\bar{x}$ of investment) ($\bar{x} = 5$)	y (deviation from $\bar{y}$ of profit) ($\bar{y} = 10$)	x^2	y^2	xy
1	3	8	−2	−2	4	4	4
2	4	5	−1	−5	1	25	5
3	6	7	+1	−3	1	9	3
4	7	20	+2	+10	4	100	20
	20	40			10	138	32

Table 10.2 shows the investment and profit of a company and the data needed to calculate the product moment coefficient:

$$\bar{x}\,(\text{investment}) = \frac{20}{4} = 5, \quad \bar{y}\,(\text{profit}) = \frac{40}{4} = 10$$

$$\Sigma(xy) = 32, \quad \Sigma x^2 = 10, \quad \Sigma y^2 = 138$$

$$r = \frac{32}{\sqrt{10 \times 138}} = \frac{32}{37.15} = \mathbf{0.86}$$

In Table 10.2 the coefficient (+0.86) shows a strong positive correlation between the two variables.

THE MEANING OF r

Direction
The sign + or − indicates the direction of association (positive or negative) between the two variables.

Strength
The numerical value of r indicates the strength of the linear relationship. r can only lie between −1 and +1: +0.9 indicates strong positive correlation, −0.9 indicates strong negative correlation, and + or −0.1 indicates weak correlation. Below about 0.7 correlation is not very strong (r^2 would be 0.49, see Variability below). The importance of differences between say 0.9 and 0.8 should not be over-emphasised. All that can be said is that they both indicate strong correlation.

Variability

r is the proportion of the variability in one variable that can be accounted for by its linear relationship with the other variable. Therefore, if $r = +0.8$ then $r^2 = +0.64$. It can be said that 64% of the variability in one variable (say x) can be accounted for by its linear relationship with the other variable (y).

Reliability

When the number of observations is very small it is not possible to place any reliance on the value of r, because any apparent relationship between the two variables may be by chance or coincidence. With a large number of observations this will be less likely.

Lag

Although strong correlation may be shown between two variables, there may be a lapse of time (a time lag) before one item influences the other. For instance advertising may not have an immediate effect on sales. This time lag may give an impression of negative correlation, when in fact correlation would be positive if due allowance was made for lag.

10.5 Coefficient of rank correlation

Some information is provided only on an ordinal scale; other information may be on an interval scale but is ordered or ranked.

> **RANKING**
>
> The capacity of containers such as bottles and boxes may be known, but they may be listed in size order from the smallest to the largest without the detail of their capacity included.

In cases of ranking it is possible to calculate a coefficient of rank correlation (r'). The interpretation of r' is the same as for r.

To calculate the coefficient of rank correlation, follow this procedure.

1   Rank each variable in order.
2   Put the corresponding rank of one variable against the rank of the other variable.
3   Show the difference between the rankings in each case.
4   Square the differences and find the sum of the squares.
5   Apply the formula:

$$r' = 1 - \frac{6\Sigma d^2}{n(n^2 - 1)}$$

where d^2 is the square of the differences or deviations in the rankings
        n is the number of units

Table 10.3 Rank correlation coefficient.

Commodity	(1) Predicted rankings	(2) Test rankings	(3) $\dfrac{d}{(2)-(1)}$	(4) $\dfrac{d^2}{(3)^2}$
A	1	5	4	16
B	2	3	1	1
C	3	4	1	1
D	4	2	2	4
E	5	1	4	16
				$\Sigma d^2 = 38$

Test results on a commodity range for the effects of weathering are produced in ranked order. These results are compared with the predicted rankings based on the judgement of a group of experts. Both rankings are shown in Table 10.3.

Applying the formula

$$r' = 1 - \frac{6 \sum d^2}{n(n^2 - 1)}$$

to the data in Table 10.3, we have:

$$r' = 1 - \frac{6 \times 38}{5(25 - 1)}$$

$$= 1 - \frac{228}{5 \times 24}$$

$$= 1 - \frac{228}{120}$$

$$= 1 - 1.9$$
$$= -0.9$$

This indicates a high level of negative correlation between the tests and the predicted results.

10.6 Spurious correlation

It cannot be over-emphasised that because two variables are associated and may produce a high correlation coefficient, this does not prove that there is a causal relationship between them. **Correlation does not indicate cause and effect.**

CAUSE AND EFFECT

There may be a strong positive correlation between sales revenue and advertising expenditure over a given period of time. It may be assumed that high sales have been caused by greater advertising expenditure. However, this is not the only possible interpretation. It could be that in years when sales were high and the company was prosperous a higher advertising budget was allocated. Or possibly, both higher sales and greater advertising expenditure arose from a general expansion in the economy.

There may be no direct connection between two variables that produce a high correlation coefficient.

INDIRECT CONNECTIONS

Sales of computers may show a high positive correlation with the level of television viewing. There may be some connection between them because both computers and television can be described as being part of the electronics industry, but the question in correlation is whether one directly influences the other. It seems unlikely that increased sales of computers will increase television viewing directly, or that if television viewing was reduced this would necessarily reduce the sale of computers. Therefore it may be that, although the correlation coefficient between the two variables is high, there is in fact very limited association between them.

It is possible to want to be able to predict the occurrence of a variable and therefore to search among a mass of data to find anything which correlates highly with this variable.

COINCIDENCE

At various times in history economic recession has been blamed on 'sunspots', because it has been noticed that these have appeared during recessions. Therefore it has been argued that there is a strong positive correlation between the appearance of sunspots and the probability of recession. It can be argued that sunspots may be connected in some way with adverse weather conditions causing poor harvests and therefore encouraging recession. However, it is generally felt (by economists and agricultural experts) that this has not been well established and is an example of spurious correlation.

Correlation is no more than an indicator of possible association between two variables. It is one piece of evidence which can lead to further investigation. Therefore it can be seen as a first stage. If little association is indicated by a correlation coefficient or a scatter diagram, then there is little point in applying the next technique, which is regression.

10.7 Regression

Regression attempts to show the relationship between two variables by providing a mean line which best indicates the trend of the points or co-ordinates on a graph.

Correlation coefficients do not provide any information on the slope of the line between two variables. **The slope indicates the rate of change in one variable against the other.** Regression lines do show this slope.

The line of best fit on a scatter diagram is dependent on the subjective judgement of the person who draws it. A regression line is drawn mathematically and therefore it is independent of individual judgement. The aim is to minimise the total divergence of the points or co-ordinates from the line. Mathematically it has been found that the best line is one that minimises the total of the squared deviations. This is known as the method of least squares.

The method of calculating a regression line by least squares is described in Section 11.5. These aspects of correlation and regression have been included in Chapter 11 because of the close links with charting time series and trends.

ASSIGNMENTS

1 Look at the following data:

Year	Sales (£000s)	Advertising cost (£00s)
1	10	2
2	15	3
3	12	5
4	18	10
5	20	12

  (i) Calculate the product moment coefficient and a rank correlation coefficient for the above data.
  (ii) Comment on the meaning of the results.
  (iii) What are the disadvantages of using a rank correlation coefficient compared with the product moment coefficient?

2 What are the problems of interpreting the results of the calculation of a correlation coefficient?

3 Draw scatter diagrams for the information in Table 10.2 and Table 10.3. Comment on the results.

4 Discuss the meaning of the term correlation. Comment on the various methods of indicating correlation.

5 What is meant by 'spurious correlation'? Discuss the extent to which there is a connection between spurious correlation and spurious accuracy.

11 Trends and forecasting

OBJECTIVES

- To analyse the use of trends, forecasting and time series
- To provide an understanding of the calculation of moving averages, seasonal variations and irregular fluctuations
- To provide an understanding of linear trends and the calculation of three-point linear regression, and the least-squares method

11.1 Trends and forecasting

Forecasts are based on information about the way in which variables have been behaving in the past. **In forecasting it has to be assumed that the behavioural patterns that have been traced in the past will continue in the future for a reasonable time.**

What is reasonable depends on the variable. National economic prosperity or depression is likely to change only over a period of years; while the sale of ice-cream may change weekly or daily, depending on the weather. Yet even with ice-cream there are trends in sales which can be seen over much longer periods such as seasons, or over a period of years.

Statistical projections do not necessarily produce accurate forecasts, because any analysis of trends depends on the assumption of stable political, economic and social conditions. However, trends can provide the possibility of the following.

- **Control:** for planning it is useful to monitor what has happened and to know when circumstances change. In business and for management purposes the analysis of trends is used for such areas as budgetary control, stock-holding, investment planning and market research. National governments have similar interests in planning and control to monitor changes in productivity, the balance of payments, levels of unemployment and levels of inflation. Economic indicators, such as the Index of Retail Prices, are carefully watched for short- and long-term developments.

- **Interpolation:** this involves finding a value within the past trend which may be a useful guide to future action.
- **Extrapolation:** extending a trend into the future may provide an indication of what is likely to happen. However, this process of extrapolation always requires great caution because the variables themselves may change, circumstances may alter and the estimated trend may not be very accurate.

Statistical techniques are useful in forecasting. Averages can provide a starting-point, dispersion can show how representative an average is, correlation can indicate the level of association between sets of data, and index numbers can be used to compare present and past performance.

What is required also is an analysis of data over time in a more general application than that supplied by index numbers. This analysis is provided by a time series.

11.2 Time series

A time series consists of numerical data recorded at intervals of time: for example, weekly output, monthly sales, annual profit.

A time series is a special case of the two-variable situation found in correlation, in which one variable is always time. Time is the independent variable (graphed on the horizontal or x axis) because it changes at regular intervals (weeks, months, quarters, years). A graph is the most popular form of presentation because this illustrates most clearly the relationship between the dependent variable and time, as well as clearly showing the trend.

There is a variety of statistical techniques aimed at separating the various elements of the trend. Some of the basic techniques are covered in this chapter. The objective of analysing the trend is to learn about the behaviour of the series and to use this knowledge as a basis for future action.

Most time series can be separated into fairly clear types of trend:

- the long-term trend (also called the basic or secular trend);
- cyclical fluctuations, which are superimposed on the long-term trend;
- seasonal variations (or short-period movements);
- irregular fluctuations (or non-recurring, random, spasmodic or residual fluctuations).

The analysis of time series is aimed at disentangling these fluctuations. The techniques used to analyse these trends include:

- moving averages, to show the long-term trend;
- seasonal adjustments, to provide control for seasonal variations;
- straight-line graphs to emphasise the general direction of the trend and to remove short-term and cyclical fluctuations.

11.3 Moving averages

Moving averages are a method of repeatedly calculating a series of different average values along a time series to produce a trend line. The line produced by charting a moving average on a graph is not a straight line, but it does even out short-term and cyclical fluctuations to some extent.

Some figures will be above the average, others below it; by using an average, fluctuations are offset one against another to produce the trend. Normally the length of the moving average (two-year, five-year, etc.) will be based on the period of time (weeks, months, years) between successive peaks or successive troughs.

Table 11.1 shows the annual sales of company with a five-year moving average. This moving average is calculated as follows.

1 Look at the sales and notice that the intervals between high points and the

Table 11.1 Moving average: annual sales of company *A* over a 15-year period.

Year	Sales (£million)		Five-year moving average
1	5		
2	2		
3	4	$32 \div 5 = 6.4$ 6.4	6.4
4	9		8.8
5	12		10.4
6	17		10.8
7	10		11.6
8	6		13.2
9	13		13.4
10	20		13.2
11	18		15.2
12	9		17.0
13	16		17.0
14	22		
15	20		

intervals between low points are about five years. Therefore a five-year moving average is calculated.

2   Add up the sales figures for the first five years and divide by 5 to produce an average ($32 \div 5 = 6.4$).

3   Subtract the sales for the first year and add the sales for the sixth year to the sales for the first five years and calculate the next moving average figure ($32 - 5 + 17 = 32 + 12 = 44$. Average is $44 \div 5 = 8.8$).

4   Continue this through the series.

5   Plot the moving average trend line against the sales figures, as shown in Figure 11.1.

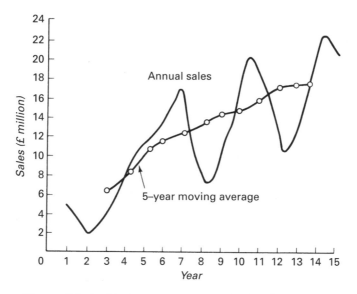

Figure 11.1   *Moving average: annual sales of company A*

From Figure 11.1 it is clear that:

- the long-term trend of sales is upwards,
- there are cyclical fluctuations at fairly regular intervals.

The moving average evens out the cyclical fluctuations and provides a clear indication of the long-term trend. More data would be required to extend the moving average across the whole series.

11.4 Seasonal variations

The effects of seasonal variations are often eliminated from data in order to clarify the underlying trend.

There are a number of methods of adjusting for seasonal variations. Table 11.2 shows a commonly used method which illustrates the general principles involved.

Table 11.2 Seasonal adjustments.

Year and quarter	(1) Sales (£million)	(2) 4-quarterly totals	(3) Centred totals	(4) Trend	(5) Variations from the trend
1 a	11				
b	8	50			
c	13	54	104	13	0
d	18	55	109	13.6	+4.4
2 a	15	60	115	14.3	+0.7
b	9	66	126	15.8	−6.8
c	18	67	133	16.6	+1.4
d	24	69	136	17	+7
3 a	16	76	145	18.1	−2.1
b	11	80	156	19.5	−8.5
c	25	82	162	20.3	+4.7
d	28	83	165	20.6	+7.4
4 a	18	82	165	20.6	−2.6
b	12	84	166	20.8	−8.8
c	24				
d	30				

In Table 11.2:

- column (1) shows the sales in £million,
- column (2) shows the 4-quarterly totals (11 + 8 + 13 + 18 = 50, 8 + 13 + 18 + 15 = 54, etc.),
- column (3) shows the total of each pair of 4-quarterly totals (50 + 54 = 104, 54 + 55 = 109, etc.),

- column (4) shows the trend, arrived at by dividing the centred total in column (3) by 8 (104 ÷ 8 = 13, 109 ÷ 8 = 13.6, etc.),
- column (5) shows the variations between sales in column (1) and the trend in column (4) (13 − 13 = 0, 18 − 13.6 = 4.4, etc.).

It is from column (5) of Table 11.2 that the seasonal variations are calculated (Table 11.3).

Table 11.3 Seasonal variations.

| | Quarter | | | |
	a	b	c	d
Year 1	−	−	0	+4.4
Year 2	+0.7	−6.8	+1.4	+7
Year 3	−2.1	−8.5	+4.7	+7.4
Year 4	−2.6	−8.8		
Totals	−4	−24.1	+6.1	+18.8
(1) Average	−1.3	−8.03	+2.03	+6.27
(2) Adjustment	+0.25	+0.25	+0.25	+0.25
(3) Seasonal variation	−1.05	−7.8	+2.3	+6.5

Table 11.3 can be explained as follows.

- For each of the quarters, the average (1) is the total divided by the appropriate number of quarters: (+0.7 − 2.1 − 2.6) ÷ 3 = −4 ÷ 3 = −1.3.
- The adjustment (2) is based on the 'unexplained' deviations that have not been entirely eliminated. If they had, the sum of the averages would have been zero: −1.3 − 8.03 + 2.03 + 6.27 = −9.33 + 8.3 = −1 (or −1.03).
- This excess negative figure is taken from each quarter equally by dividing by 4 (1 ÷ 4 = 0.25) and adding this figure (0.25) to each quarterly average. It is added to each quarterly figure because it is not clear where the difference arises. In this case the difference is very small, but it can be large in some cases.
- The final row (3) is the estimate of seasonal variations for each quarter. The figures are rounded to one decimal place, because to extend the figures further would indicate a higher degree of accuracy than is likely.

These figures can be used to produce a seasonally adjusted series and they can be applied to future years (Table 11.4 and Figure 11.2).

Notice that in Table 11.4 the minus figures for seasonal variation have

Table 11.4 Seasonally adjusted sales figures: year 5 sales figures for company *B*.

Quarter	Sales (£million)	Seasonal adjustment	Seasonally adjusted sales (£million)
a	20	+1.05	21.1
b	15	+7.8	22.8
c	25	-2.3	22.7
d	35	-6.5	28.5

Figure 11.2 Seasonally adjusted sales figures: year 5 sales figures for company B

been added and the plus figures subtracted. What is being done is to eliminate seasonal variation and therefore additions are made to the generally low seasons, and subtractions made to the seasons in which the variable has a high value.

It is clear from Figure 11.2 that when seasonal variations are eliminated the general trend of sales is upwards although not quite as sharply as indicated by the fourth quarter's unadjusted results. The fall in sales in the second quarter appears to be because of seasonal factors.

These points are borne out if the whole series is drawn on a time series graph (Table 11.5 and Figure 11.3).

Table 11.5   Time series with seasonal adjustments.

Year and quarter		Sales (£million)	Seasonally adjusted sales (£million)
1	a	11	12.1
	b	8	15.8
	c	13	10.7
	d	18	11.5
2	a	15	16.1
	b	9	16.8
	c	18	15.7
	d	24	17.5
3	a	16	17.1
	b	11	18.8
	c	25	22.7
	d	28	21.5
4	a	18	19.1
	b	12	19.8
	c	24	21.7
	d	30	23.5

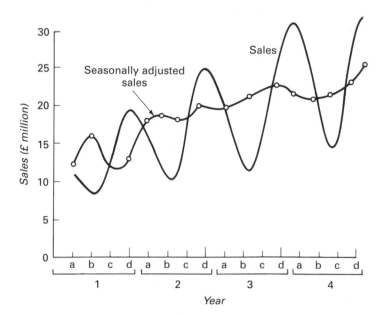

Figure 11.3   Time series with seasonal adjustments

11.5 Irregular and residual fluctuations

In a time series the curve of the graph reflects the trend, seasonal variations and irregular factors. The value of the residuals can be calculated once the trend and the seasonal variations have been calculated (Table 11.6).

In Table 11.6 a residual figure has been calculated in the last column. The value of calculating this figure is to give an indication of the extent to which the series has been affected by irregular external factors. The residual values calculated in Table 11.6 are small and therefore any forecast made on the basis of the time series would be unlikely to be upset by external factors. If the residual factors had been large then any forecasts made would be less reliable.

Table 11.6 Residual factors.

Year and quarter		Original series	=	Trend	+	Seasonal variation	+	Residual
1	a	11						
	b	8						
	c	13		13		+2.3		−2.3
	d	18		13.6		+6.5		−2.1
2	a	15		14.3		−1		+1.7
	b	9		15.8		−7.8		+1
	c	18		16.6		+2.3		−0.9
	d	24		17		+6.5		+0.5
3	a	16		18.1		−1		−1.1
	b	11		19.5		−7.8		−0.7
	c	25		20.3		+2.3		+2.4
	d	28		20.6		+6.5		+0.9
4	a	18		20.6		−1		−1.6
	b	12		20.8		−7.8		−1
	c	24						
	d	30						

11.6 Linear trends

Trends can be shown by a straight line if there is a linear relationship between the variables. The straight line can be based on a mathematical equation, such as with **the least-squares method,** or the straight line can be drawn by eye to provide a **'line of best fit'** (as for scatter diagrams, see Chapter 10). **Where there is a strong, linear correlation between two variables it is possible to use the method of** *three-point or two-point linear estimation* (also known as linear

Table 11.7   Three-point linear estimation.

Year	Sales (£million)		
1	5		
2	2		
3	4	Average sales = $\frac{59}{7}$ = £8.4	(iii)
4	9	Central year = 4	
5	12		
6	17		
7	10	Arithmetic mean = $\frac{183}{15}$ = £12.2	(i)
8	6	Central year = 8	
9	13		
10	20		
11	18	Average sales = $\frac{118}{7}$ = £16.9	(ii)
12	9	Central year = 12	
13	16		
14	22		
15	20		

regression and the method of partial averages) **to produce a straight line or an approximation to it.**

Three-point linear estimation

1   Calculate the arithmetic mean of each variable: this provides the co-ordinates for point (i) (see Table 11.7 and Figure 11.4).
2   Calculate the arithmetic mean of the figures which are positioned above point (i) for each variable: this provides the co-ordinates for point (ii).
3   Calculate the arithmetic mean of the figures which are positioned below point (i) for each variable: this provides the co-ordinates for point (iii).
4   These three points are plotted on a graph to provide the trend line, the line is drawn through the overall mean values and as nearly as possible through the other two points (see Table 11.7 and Figure 11.4).

Two-point linear estimation

1   Divide the series into two equal halves, each containing a complete number of years (if the series contains an odd number of years, omit the middle year).
2   Calculate the mean of each half.
3   Plot these two mean halves on a graph and join them to produce a trend line.

In the example in Table 11.7 the two means would be £8.4 and £16.9. These two points could be plotted and joined to form the trend line.

This method is not very accurate if there are any extreme values in the

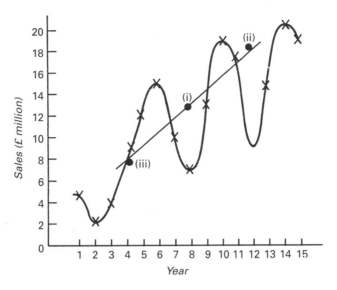

Figure 11.4 Three-point linear estimation

series. Both methods are approximations to a mathematically produced linear trend, but are likely to be more accurate than a line of best fit drawn by eye.

The least-squares method

The aim is to produce a line which minimises all positive and negative deviations of the data from a straight line drawn through the data. This is carried out by squaring the deviations (to remove minus items) and therefore the least squares is the 'best' line, the line which minimises the error in the direction of the variable being predicted.

The straight-line equation

$$y = a + bx$$

can be used where:

y and x are variables
a represents the intercept
b represents the slope or gradient of the line (see Figure 11.5).

Where the equations for finding the slope and the intercept are:

$$b = \frac{\Sigma xy - \dfrac{\Sigma x \times \Sigma y}{n}}{\Sigma x^2 - \dfrac{(\Sigma x)^2}{n}} \quad \text{or} \quad \frac{n\Sigma xy - \Sigma x\Sigma y}{n\Sigma x^2 - (\Sigma x)^2}$$

$$a = \bar{y} - b\bar{x}$$

the straight-line equation $y = a + bx$ can be rewritten as:

$$y = \bar{y} - b\bar{x} + bx$$

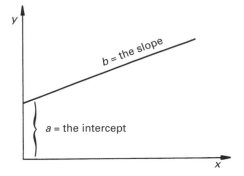

Figure 11.5 The slope and the intercept

where $\bar{y}$ and $\bar{x}$ are the arithmetic means of the two variables. This equation, in the same way as the previous one, minimises the positive and negative deviations of the data from a straight-line trend.

In a time series one of the two sets of data (x) is time and this increases by equal amounts (months or years). This time series does not have a 'value', but it is a series so that it is possible to number the years (months, week or days) in order as an increase of 1 on the last (0, 1, 2, 3, etc.). When the variable x is not time, then this line is a regression line.

An example of the calculation of the least-squares method is shown in Table 11.8.

Table 11.8 Least-squares method.

Year x	Sales (£million) y	x^2	y^2	xy
0	5	0	25	0
1	2	1	4	2
2	4	4	16	8
3	9	9	81	27
4	12	16	144	48
5	17	25	289	85
6	10	36	100	60
7	6	49	36	42
8	13	64	169	104
9	20	81	400	180
10	18	100	324	180
11	9	121	81	99
12	16	144	256	192
13	22	169	484	286
14	20	196	400	280
105	183	1015	2809	1593

In Table 11.8:

$$\bar{x} = \frac{105}{15} = 7 \quad \bar{y} = \frac{183}{15} = 12.2$$

$$b = \frac{\Sigma xy - \dfrac{\Sigma x \times \Sigma y}{n}}{\Sigma x^2 - \dfrac{(\Sigma x)^2}{n}}$$

$$= \frac{1593 - \dfrac{105 \times 183}{15}}{1015 - \dfrac{105^2}{15}}$$

$$= \frac{1593 - 1281}{1015 - 735}$$

$$= \frac{312}{280}$$

$$= 1.11$$

$$y = a + bx \quad \text{or} \quad y = \bar{y} - b\bar{x} + bx$$
$$y = 12.2 - (1.11 \times 7) + (1.11 \times x)$$
$$= 12.2 - 7.77 + 1.11x$$
$$= \mathbf{4.43 + 1.11x}$$

Therefore $y = 4.43 + 1.11x$.

This means that each year (x) there will be an average increase of £1.11 (million) in sales. When $x = 5$, y will equal $4.43 + 1.11 \times 5 - 9.98$.

If this is applied to the annual sales of company A, the results in Table 11.9 are produced. The linear (least-squares) trend in Table 11.9 can be graphed against sales (Figure 11.6); see page 184.

This least-squares method of finding a linear trend is based on averaging out deviations. The straight line suggests a limited set of influences acting together in a single direction. It smoothes out short-term and cyclical fluctuations to emphasise the basic trend.

11.7 Conclusions

The methods for finding trends described in this chapter are only some of the many available. This is an area in which there has been a good deal of research and development because of the desire by business and government to be able to forecast. Analysing the various types of trend can indicate a pattern in the data which may be repeated in the future. However, this analysis is not a substitute for looking carefully at the original figures and attempting to analyse why poor results occurred in some years and good results in others. It may be possible to

Table 11.9 Linear trend.

Year	Sales (£million)	Linear trend
0	5	$y = 4.43 + 1.11 \times \ 0 = \ 4.43$
1	2	$y = 4.43 + 1.11 \times \ 1 = \ 5.54$
2	4	$y = 4.43 + 1.11 \times \ 2 = \ 6.65$
3	9	$y = 4.43 + 1.11 \times \ 3 = \ 7.76$
4	12	$y = 4.43 + 1.11 \times \ 4 = \ 8.87$
5	17	$y = 4.43 + 1.11 \times \ 5 = \ 9.98$
6	10	$y = 4.43 + 1.11 \times \ 6 = 11.09$
7	6	$y = 4.43 + 1.11 \times \ 7 = 12.20$
8	13	$y = 4.43 + 1.11 \times \ 8 = 13.31$
9	20	$y = 4.43 + 1.11 \times \ 9 = 14.42$
10	18	$y = 4.43 + 1.11 \times 10 = 15.53$
11	9	$y = 4.43 + 1.11 \times 11 = 16.64$
12	16	$y = 4.43 + 1.11 \times 12 = 17.75$
13	22	$y = 4.43 + 1.11 \times 13 = 18.86$
14	20	$y = 4.43 + 1.11 \times 14 = 19.97$

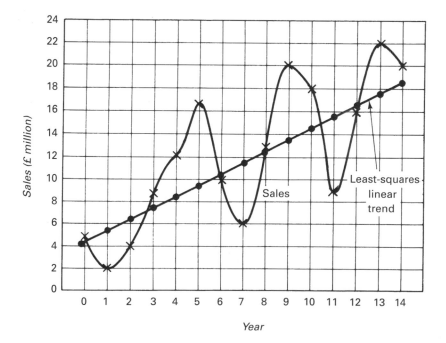

Figure 11.6 The slope and the intercept

avoid repeating the poor results and to try to ensure the repetition of the good results.

Any forecasts that are made need to be surrounded with qualifications because of all the unexpected and unforeseen variables that can influence future performance. It is still well worth forecasting because planning can provide control over events and avoid constant last-minute problem solving.

Statistics can provide the basis for planning, decision making and forecasting. The methods and concepts outlined in this book are aimed at providing a framework which can be used for understanding, collecting, presenting, summarising, comparing and interpreting statistics. It is only by 'handling' statistics, that is with practice, that confidence in the ability to use these techniques can be created.

ASSIGNMENTS

1   Look at the table below showing company sales (£000s) over five years.

| | Quarter | | | |
	a	b	c	d
Year 1			120	100
Year 2	85	140	130	90
Year 3	90	120	125	85
Year 4	110	125	140	95
Year 5	115	130		

Show the trend for the above time series by calculating a moving average. Plot the sales and the moving average on a graph. Calculate the seasonal variations for these figures and show the seasonally adjusted sales curve on the graph. Comment on your results and on the three curves on the graph.

2   Calculate a three-point linear trend line from the weekly wage figures in Table 7.1. Calculate a trend line by the least-squares method. Draw a graph showing the two lines. Comment on the validity of the methods of producing a linear trend and on the information brought out by the lines.

3   Discuss the reasons for attempting to forecast trends and the problems of statistical forecasting.

4   Find details of forecasting made by the government, companies and other organisations. Consider the basis on which they have been made and, if possible, how accurate they have proved to be.

5   Consider in detail the problems of weather forecasting. How applicable are these problems to the problems of forecasting faced by business?

⬡12 Mastering more statistics

OBJECTIVES

- To provide an understanding of statistical proof, experiments and inductive statistics
- To consider further methods of collecting information, including market research, as well as checklists, ratings and scaling
- To consider the use of information technology in statistics

12.1 More statistics

The purpose of this book is to provide a comprehensive introduction to statistics, an appreciation of the subject and a foundation for further study. It is intended to demonstrate the practical applications of statistical techniques. The next step is to progress further along the statistical road into particular areas of specialisation and interest.

These areas will depend very much on the individual and the circumstances involved. The social scientist, for example, may want to look more closely at the application of statistics to a particular subject or topic, while the student of business studies might be particularly interested in the application of statistics to market research or information technology, and the businessperson might be interested mainly in statistical sources of information, the interpretation of data and forecasting. This book should provide a stepping-off point for all these interests.

This chapter will provide a brief introduction to some areas of specialisation and interest in order to point the way forward.

12.2 Statistical proof

One area of further study is a 'step back' to look at the proof of the formulae and the mathematical basis for the techniques used here. This requires a greater mathematical and algebraic knowledge than is assumed in this book; it is possible, however, to provide an insight into this area through simple examples.

'x' notation

The formula for the arithmetic mean used in Chapter 6 is:

$$\bar{x} = \frac{\Sigma x}{n} \quad \text{where } x = \text{the value of the items}$$

$$n = \text{the number of items}$$

This is arrived at by taking the score of each individual or item as x. x_1 represents the score of the first individual, x_2 that of the second and so on to x_n (the number of individuals or items included). The formula can be written in these terms, that is, in the form of x notation:

$$\left[\bar{x} = \frac{x_1 + x_2 + \ldots + x_n}{n} \right]$$

To take this a little further, note that the mean has a number of properties or characteristics, one of which is that the sum of the deviations of each score from the mean will always be zero (see Section 6.2). This can be expressed symbolically by the equation:

$$[\Sigma(x_i - \bar{x}) = 0]$$

when x_i represents the score of the general individual or item.

The proof of this property is that the above equation expresses a sum of numbers each of which is actually a difference and the expression can be broken into the differences of the two sums:

$$[\Sigma(x_i - \bar{x}) = \Sigma x_i - \Sigma \bar{x}]$$

Σx_i and $\Sigma \bar{x}$ are equal to each other, $\bar{x}$ is a constant and this means that:

$$\left[\Sigma \bar{x} = n\bar{x} = n\frac{\Sigma x_i}{n} = \Sigma x_i \right]$$

The difference between Σx_i and $\Sigma \bar{x}$ is zero.

This property can be used to simplify the calculation of the mean by using the assumed mean (see Section 6.2). The formula

$$\left[\bar{x} = x \pm \frac{\Sigma d_x}{n} \right]$$

can be rewritten as

$$\left[\bar{x} = \bar{x}' \pm \frac{\Sigma(x_i - \bar{x})}{n}\right]$$

where $\bar{x}'$ is the assumed mean

$\Sigma(x_i - \bar{x})$ is the sum of the deviations from the assumed mean

n is the number of individuals or items.

'x' NOTATION

Six people each earn £80, £100, £108, £160, £180, £212 respectively. For the purpose of calculation the average earnings can be assumed to be £150.

Item (£)	Deviation from assumed mean (£150) $(x_i - \bar{x})$	
80	−70	
100	−50	
108	−42	
160		+10
180		+30
212		+62
	−162	+102 $\Sigma(x_i - \bar{x}) = -60$

$$\bar{x} = \bar{x}' \pm \frac{\Sigma(x_i - \bar{x})}{n}$$

$$= £150 - \frac{60}{6}$$

$$= £150 - 10$$

$$= £140 \qquad \text{Average earnings} = £140$$

Subscripts

Subscripts are notations above and below a symbol which are used to clarify its use. Where there are no ambiguities, subscripts can be dropped from formulae.

Σ is used to indicate a summation, so that all quantities appearing to the right of it should be summed. Instead of using completely different letters for each quantity being summed (for example, $a, b, c, d, e, \ldots$) a single letter is used (such as x or y) which represents a range of numerical values $(x_1, x_2, x_3, \ldots)$.

Σ can be used as follows:

$$\left[\sum_{i=1}^{n} x_i = x_1 + x_2 + x_3 + \ldots + x_n\right]$$

The notations above and below Σ are used to indicate that i takes on the successive values 1, 2, 3 up to n.

Similarly it would be possible to write:

$$\sum_{i=2}^{5} x_i = x_2 + x_3 + x_4 + x_5$$

where the scores of the second to the fifth observation are to be added together.

Use of 'x' notation and subscripts

It is not essential to use 'x' notation or subscripts if the aim is to calculate and use a statistic such as the arithmetic mean. It is essential, however, to use these techniques for a conceptual understanding of the subject and to acquire the ability to adapt and develop formulae. This can be compared perhaps, with the difference between on the one hand following instructions in a cookery book to produce a successful dish by mixing ingredients described in a recipe in the correct quantities, order and temperature, and on the other hand having the knowledge and confidence to throw the cookery book away and create individualised and new dishes.

12.3 Experiments

Another area of further study is to look more closely at survey methods and to consider attempts to establish causal relationships between variables. Many surveys are designed to measure certain characteristics of the population, for example spending habits, while another type of survey is concerned with possible causal connections between variables, such as the effect of advertising on spending habits. This is a more complex type of survey than the descriptive type and can be classified as a form of experiment.

The basis of experiments

An experiment involves a degree of planning and control by the experimenter. In the 'ideal' case of the laboratory experiment, the environment is controlled and the effect of one variable on another can be studied in isolation from extraneous influences. In physical sciences, in contrast to human and social sciences, the dependence of one measurable quality on another can be expressed in exact terms, such as volume or temperature. In social research, experimentation may be impossible and causal influences may have to be based on non-experimental studies.

> **An experiment involves the manipulation of the independent variable (x) to see what effect this may have on the dependent variable (y).** The social scientist uses the same techniques as the physical or natural scientist. Phenomena are carefully measured and the data used to test hypotheses which attempt to explain what is happening. The results of experiments can be used to make statements about general tendencies or 'laws' which can be used for prediction.

In statistical experiments randomisation can be used to reduce chance factors. Comparisons can be made between two groups of people or items by matching for extraneous variables. Matching is necessary because comparison between two or more groups may be complicated by the fact that they differ on a third variable (z). This difference may explain the association between the independent and dependent variables. If however the two groups can be made equivalent with regard to z, then z cannot be the explanation for the association. This can be achieved by 'matching' techniques. For example, if two groups of people vary in age or sex this may explain the difference in, say, attitude between them. The two groups can be 'matched' so that the age and sex distributions are the same.

Matching is difficult when there are a number of variables to control for and 'randomisation' can be used in these circumstances. If subjects can be randomly assigned to the experimental and control groups then the differences between the two groups for any prior extraneous variable can be only a random chance fluctuation and statistical techniques can be used to determine the probability that such random fluctuations could lead to an association (see Chapter 8). Randomisation is not always possible with human populations, but the fact that the social scientist's measuring and testing does not take place in a laboratory does not in itself make it any less 'scientific' but does make it difficult to arrive at definite answers. This is particularly the case in behavioural surveys because people's behaviour is unpredictable.

The validity of experiments

In order to make an assessment of a causal relationship it is necessary to know not only of its existence but also of the degree of association. There is a considerable difference, for example, between a small change in consumer spending on a commodity as a result of an advertising campaign and a large increase in sales of the commodity. The time sequence of the variables also provides an important factor in determining causality. If y precedes x, x cannot be a cause of y, so that, for example, if sales increase before the marketing campaign, then the campaign could not have caused the increased sales.

At the same time it is important to rule out other variables as the explanation of the association. It might be a third variable z which is the cause of x and y, and it is only for this reason that they are associated. This does not make their association any less real, but any explanation which omits z is spurious. To investigate whether x (the independent variable) has any effect on y (the dependent variable) a study to measure association is needed and a number of experimental designs are used to do this.

EXPERIMENTAL DESIGNS

The 'after-only' design

The 'after-only' design is the simplest design; the effect on y is assessed by measuring its incidence in both the experimental group and the comparable control group after the former has been exposed to the independent variable x. For example, the experimental group might be exposed to a marketing

---

campaign of which the control group is kept ignorant. If the attitude of the two groups to the object of the campaign is found to be similar then the marketing campaign can be considered to be ineffective.

The 'before-after' design

The 'before-after' design uses the same individuals for the experimental and control group, measuring y before and after the group has been exposed to x. For example, if the aim is to evaluate the effectiveness of a marketing video in changing attitudes towards a product, the change, if any, in score on an attitude scale (or inventory) given before and after seeing the video might indicate its effectiveness. This is based on the assumption that without the video the attitude would not have changed between the two measurements.

The 'before-after' design, with a 'control' group

In this design both the experimental and control groups are measured before and after the experimental group is exposed to x (say, the video film). The effect of x is then measured by comparing the changes in y (the attitude) in the two groups. This design reduces the probability of 'other explanations' which are always a problem in drawing conclusions about causation, because it is possible that another factor may explain the change in y and thereby invalidate the conclusion that x caused the change.

'Other explanations'

There is a range of explanations for a causal association between x and y other than the explanation that is the object of the experiment. These include the following.

- **'History':** this involves such material and events as newspaper reports, personal incidents or other factors which cause a change in attitude.
- **'Maturation':** the effects of time and changing experience may alter opinions and attitudes, even over a few hours.
- **Testing:** in an experimental design which includes a pre-test and a post-test, the experience gained through the pre-test may result in an increase in the post-test scores.
- **Instrumentation:** if two tests are given (before and after), they may not be exactly comparable (and, of course, the same test cannot be given twice) and this can lead to differences in score.
- **Statistical regression:** if individuals are chosen for the experimental group on the basis of extreme scores on y, it has been observed that they often produce an average score on the post or 'after' test closer to the population average. This result tends to occur even if any exposure to x did not have an effect, it can arise as a result of chance factors and measurement errors.
- **Selection:** the experimental and control groups must be as comparable as possible and if the two groups are too different the results may be invalidated. An advertising campaign, for example, may be tested on an experimental group in Brighton and a control group in Liverpool. It could

be argued that there are many differences between the two areas which would provide an alternative hypothesis for any difference observed.

- **Experimental 'mortality':** comparability may be weakened by failure to obtain the required measurements for every member of each group. This can arise if experiments take place over a period of time during which individuals may become hard to contact because, for example, they have moved jobs or houses.

- **Interactive and reactive effects of selection, testing and experimental arrangements:** testing may sensitise a group to a particular topic and the 'Hawthorne' effect may develop. The Hawthorne studies enquired into the relationship between physical work conditions and productivity. The output of a group of workers was studied under various conditions. The group was treated differently from usual and knew they were the subject of an experiment. The effect of these arrangements was found to be so strong as to hide any effects of changes in physical conditions.

In experimental and survey design there is a difference between internal and external validity. Internal validity is concerned with the question of whether a true measure of the effect of x is obtained from the subjects of the experiment, while external validity is concerned with the generalisations of the findings to other people. The demands of these two forms of validity compete because the stronger the design is made in internal validity, the weaker it becomes in external validity. This means that, in general, surveys are strong on external validity so that their results can be generalised to other people, although their design may be questioned; with experiments it is the other way around. Surveys will tend to be used when there is a generous margin of error and the results need to be widely applicable, for example, for opinion polls and aspects of market research. Experiments will be used where the results need to be very accurate and the area of study is relatively narrow, for example, whether a particular aspect of a marketing campaign is effective.

12.4 More inductive statistics

Chapter 8 provides an introduction to statistical induction as the process of drawing general conclusions from a study of representative cases. This is an area of considerable scope for further study. **Statistical induction involves the theories of probability and statistical estimation and includes a range of tests** because social scientists in particular are concerned with whether or not an individual or item possesses a certain attribute and whether an experiment has been a success or failure.

The binomial and poisson distributions

Whenever it is possible to hypothesise a certain probability of success, whenever trials are independent of each other and whenever the number of trials is relatively small, it is possible to make use of statistical tests. These involve probability distributions which are similar to frequency distributions except that probabilities are used instead of frequencies.

A binomial distribution is concerned with two items (hence its name), such as the probability that an event occurs and the probability that the event does not occur; while **the poisson distribution provides a good approximation to the binomial distribution.** Both distributions are discrete while the normal distribution is continuous (see Section 4.3); they can be used whenever there are events which occur in a random manner.

The sign test is used in simple before-and-after experiments when a small number of cases are used to determine whether the experiment has been successful or not. If the experiment is concerned with whether an advertisement has resulted in people changing their habits, a sign is designated for successes (say $+$) and failures (say $-$) and then, making assumptions about both the population and the method of sampling, the level of probabilities of outcomes can be assessed within critical values or a critical region.

The poisson distribution can be used when large samples are drawn from a population containing a small number of the items being considered, that is where the sample size (n) is large and the probability (p) of a success is very small. It can be used where the average number of successes is known but the sample size is not known, and in looking for defective or unusual items in large batches. The number of accidents in a particular industry, for example, can be counted over a year and used as an estimate of the average number of accidents. It is not possible, however, to count the number of times accidents have not occurred.

Statistical tests

There is a range of tests which can be used to analyse the results of a sample or experiment. Although these tests have a number of similarities, they tend to be used under different circumstances and care has to be taken to use the correct test. Detailed examples of these tests are outside the scope of this book; however, the descriptions given here provide a flavour of the range of tests available to arrive at statistical decisions.

POWER OF A TEST

The power of a test is defined as $1 - \beta$ (β = Greek beta) or $1 -$ the probability of a type II error (see Section 8.6). The power of a test is inversely related to the risk of failing to reject a false hypothesis. The greater the ability of a test to eliminate false hypotheses, the greater is its relative power.

The most common tests include the following.

Student's t distribution

This was introduced by W. S. Gossett writing under the name of Student (because the firm for which he worked did not at that time approve of members of its staff publishing papers under their own name). **The 't' test assumes a normal distribution and it is used to estimate averages when only comparatively small samples are available,** while if the sample is large t can be approximated to z (see Chapter 8).

It is when samples are small, however, that there is the most doubt about the exact nature of the population. If a sample contains 20 items it is difficult to accept the assumption of a normal distribution and this means that the t test may not be very useful. There are alternatives to the t test which do not involve this assumption (such as non-parametric tests).

The difference-of-means test

This extends the single means test to a test in which a comparison can be made between the means of two samples. The samples must be selected independently of each other so that not only must there be independence within each sample, assured by randomness, there must also be independence between samples. The sampling distribution is for a difference between sample means, the mean of the sampling distribution is given by the difference between two population means rather than either of them separately. If the means are equal so that $\mu_1 = \mu_2$, the mean of the sampling distribution will be positive. For example, if $\mu_1 = 80$ and $\mu_2 = 40$, the distribution of $\bar{x} - \bar{x}_2$ will have 40 as its mean value.

> **Non-parametric tests are tests that do not require the assumption of a normal distribution.** They are used when an interval scale cannot be used but the ordering of scores is justified and the sample is small. They include the Runs test, the Mann–Whitney or Wilcoxon test and the Kolmogorov-Smirnov test.

The Wold–Wolfowitz runs test

This uses the assumption that the level of measurement is at least an ordinal scale and the two samples have been drawn from the same continuous population or two identical populations. The data from both samples are taken and the scores are ranked from high to low, ignoring the fact that they come from two different samples. If the null hypothesis is correct, then the two samples will be mixed, so that there is not a long run of cases from the first sample followed by a run of cases from the second.

> **RUNS TEST**
>
> Two samples A and B would be expected to produce a rank order such as:
>
> [AABBAAABABBAABABBAB]
>
> rather than:
>
> [AAAAAAABAABBBBBBBBB]
>
> In order to test how well the two samples are mixed when ranked, the number of runs are sampled. In the first example there is a 'run' of two As, two Bs, three As, one B and so on. The total number of runs is 12, while in the second example, there are only four runs. If the number of runs turns out to be large, as in the first example, the two samples will be mixed so that it is not possible to reject the null hypothesis. The sampling distribution of runs can be used to establish the critical region used in rejecting the null hypothesis.

The Mann–Whitney or Wilcoxon test

This test requires exactly the same assumptions as the Runs Test and involves a similar procedure. The scores of the two samples are ranked; taking each score in the second sample, the number of scores in the first sample which have larger ranks are counted. The same process is carried out for the scores of the second sample and then the results are added to provide the statistic u. The sampling distribution of u can be obtained exactly in a small sample or it can be approximated by a normal curve in the case of larger samples. If u is either unusually small or unusually large the assumption that the two samples have been drawn from the same population is rejected.

The Kolmogorov–Smirnov test

This is another two sample non-parametric test based on the same assumptions as the previous tests. Unlike the other tests, however, this test can be used in instances where there are a large number of ties in the rankings. A number of ties may arise in research in the social sciences when variables which are ordinal scales may be grouped into a few large categories, for example, in grouping occupations.

The chi-square test

The previous tests have been concerned with the existence of relationships between two variables. **The chi-square test and *Fisher's Exact Test* indicate the strength or degree of relationship between two variables.** The chi-square test (pronounced kigh-square or kie-square, so called because it is represented by the lower case Greek letter 'chi': χ) is a general test which can be used whenever there is an attempt to evaluate whether or not frequencies which have been empirically obtained differ significantly from those which would be expected under a certain set of theoretical assumptions. The test has a number of applications, the most common of which are contingency problems in which two nominal scale variables have been cross-classified.

CHI-SQUARE

In an experiment based on consumer preferences for washing powder it may be expected (the null hypothesis) that the consumer will be unable to distinguish between washing powders if they do not know their name, the shape of the packet and other aspects of advertising and sales promotion. A sample will provide *observed* frequencies which can be compared with the *expected* frequencies. The larger the differences between observed and expected frequencies, the larger the value of chi-square.

Chi-square will be zero only when all observed and expected frequencies are identical. If the value of chi-square is larger than expected by chance, it is possible to reject the null hypothesis.

Fisher's exact test

This test, developed by R. A. Fisher, **is used under the same conditions as the chi-square,** where the number of cases is small, to provide exact rather than approximate probabilities.

The analysis of variance

Whereas in the difference-of-means test, runs test, Mann–Whitney and Smirnov tests, two samples are compared for the significance of the differences between means and proportions, the chi-square test can compare more than two samples. **The analysis of variance can also be used to test for differences among the means of more than two samples.** It represents an extension of the difference-of-means test and can be used for testing for a relationship between a nominal scale and an interval scale.

In the analysis of variance, assumptions are made of a normal distribution, independent random samples and equal population deviations. The null hypothesis will be that population means are equal, although the test involves working with variances rather than means and standard errors. Where there are three variables being compared the analysis of variance provides a single test of whether or not all three differ significantly among themselves, that is whether all could have come from the same population.

12.5 More methods of collecting information

Introduction

Information can be collected by observation, surveys, experiments, interviews and questionnaires. In the application of statistical methods in a number of areas of business, public services, and social and economic research, other techniques for collecting information are used. An example of this is in market research.

Market research

Market research is the systematic gathering, recording and analysing of data about problems relating to the marketing of goods and services. It is designed to build up a picture of the market and the forces that operate it and forms the basis of forecasting the volume and value of sales.

The firm needs to build up a 'consumer profile' so that it can modify its marketing mix in order to sell its products. It needs to know who its customers are in terms of age, sex, income, region or socio-economic class in order to understand the motivation and attitudes behind consumption. **Research findings and the informed judgement of management are backed up by statistical techniques.** Moving averages isolate random cyclical and seasonal variations from the trend; correlation analysis is used to discover the extent of the link between variables such as advertising expenditure and sales; while statistical tests and the analysis of variance can be used to analyse the results of market research surveys and experiments.

Market research may be based on available sources of data such as those included in Chapter 2 as well as specific sources and documents.

READERSHIP SURVEY

The Joint Industry Committee for National Advertising Readership Surveys (JICNARS) has been the source of information on consumer profiles and readership per copy of a range of publications including newspapers and magazines. The survey, of around 15 000 people, has been taken every six months to provide the information. It has enabled advertisers to have a clear idea of cost per reader and the market served by each publication. JICNARS has used socio-economic class gradings as the basis for the consumer profile so that advertisers have known by which class group their advertisements are likely to be seen.

Socio-economic classification plays an important part in market research because of the wish to direct sales campaigns of goods and services at clearly identified consumer targets. The approach used by the Office of Population and Census Surveys (OPCS) has been based on the 'occupation of the father' while the Institute of Practitioners in Advertising (IPA) definition has been based on the 'head of household's occupation'. The two classifications related very closely (see Table 12.1) and have been commonly used as the basis for market research

Table 12.1 Socio-economic classification.

OPCS classification		IPA definition	
Category	Description	Class	Description
I	Professional occupations	A	Higher managerial, administrative or professional
II	Intermediate occupations (including most managerial and senior administrative occupations)	B	Intermediate managerial, administrative or professional
IIIN	Skilled occupations (non-manual)	C1	Supervisory or clerical and junior managerial, administrative or professional
IIIM	Skilled occupations (manual)	C2	Skilled manual workers
IV	Partly skilled occupations	D	Semi-skilled and unskilled manual workers
V	Unskilled occupations	E	State pensioners or widows (no other earners), casual or lowest grade workers, or long-term unemployed
OTHER	Residual groups including, for example, armed forces, students and those whose occupation was inadequately described		

Source: *Social Trends*

and also for such methods as quota sampling, checklists, attitude scales and indirect techniques. These classifications indicate areas for further investigation into the foundations of methods of collecting information.

As well as the use of available resources and common methods of collecting information, there is a range of other methods that are used in market research and other areas of applied statistics. These include checklists, ratings, attitude scales, projective and indirect techniques.

12.6 Checklists and ratings

Checklists

Checklists consist of lists of terms which are understood by respondents and which can express their views more briefly and succinctly than open-ended questions.

CHECKLIST

Respondents may be asked: 'Please give your opinion on the main attributes that are important for you in choosing a new car by ticking the appropriate box in the following table.' (See Table 12.2.)

	Very important	Important	Indifferent	Undesirable
1. Size				
2. Colour				
3. Safety				
4. Economy				

Checklists tend to be crude devices and they are more precise and effective when they are constructed and used to test specific hypotheses rather than as an exploratory technique. Problems can arise because respondents may try to 'help' by attempting to produce the results which are expected. At the same time, checklists may not include a complete list of attributes or descriptive words. This means that they require careful interpretation. For Table 12.2 it can be asked: Does it include all the attributes of concern to respondents in choosing a car? Will the results predict actual behaviour in the purchase of cars? Do the results reflect attitudes in the present? Pilot survey work and market research would help to answer these questions.

Ratings

Ratings give a numerical value to some kind of judgement. Examples include marks at school and for examinations, grades in proficiency reports and assessment ranking.

Ratings are used:

- as objective assessment, for example about the qualities of a car;
- in a subjective way to indicate the attitude of the respondents;
- as self-ratings of personality traits or attitudes.

Ratings may give a spurious impression of accuracy because they provide a numerical value. Examination marks, for example, may appear to be arrived at objectively while in fact, for written exams, they involve a value judgement. At the same time ratings may suffer from the **halo effect** in which instead of giving attention separately to each item, respondents may be influenced by an overall feeling of like or dislike.

HALO EFFECT

A respondent may like a particular make of car and therefore give it high marks for performance, comfort, style and safety in comparison to another car that is disliked even if this car has some obvious advantages, such as being more economical.

Ratings may suffer also from **response set**. If the rating scales are arranged one underneath the other and always with the 'good' end on one side and the 'bad' on the other, the respondent may decide in favour or against the object of the ratings and may run down the page always checking a position on one side of the scale, without actually reading the items or giving each of them separate thought. To counteract this problem the direction of scales can be randomised so that the 'good' and the 'bad' fall sometimes at one and sometimes the other side of the scale.

12.7 Scaling

Introduction

Scaling is a method of obtaining information by the use of a scale rather than a long list of individual questions. On most subjects which are surveyed or sampled it would be possible to make a long list of relevant questions, particularly when asking about attitudes and opinions or analysing particular attributes. Scaling methods are an alternative to asking questions in other ways, by utilising simultaneously a number of observations on each respondent.

An occupational preferences rating scale may include a number of statements about occupational interests on which the respondents have to indicate a level of interest on a scale from low to high. Each individual answer may not provide very much information, but between them the answers may indicate occupational preferences. The scale might include statements such as those below.

	Very untrue	Untrue	Undecided	True	Very true
I like working with people					
I am very interested in machines					
I like to work with my hands					

Scaling is often used for attitude measurement and this is based on the assumption that there are underlying dimensions along which individual attitudes can be ranged. It is assumed that by employing one of the various attitude scaling procedures a person can be assigned a numerical score to indicate a position on a dimension of interest. The score may be on a nominal, interval or ratio scale.

A nominal scale classifies individuals into groups which differ with respect to the characteristic being scaled without there being any implication of gradation between groups. An ordinal scale ranks individuals along the continuum of the characteristic being scaled, again without any implication of distance between scale positions. An interval scale has equal units of measurement so that it is possible to interpret not only the order of scale scores but also the distance between them. The ratio scale has the properties of an interval scale plus a fixed origin so that it is possible to compare both differences in scores and the relative magnitude of scores.

General procedures in scaling

There is a general procedure in producing a scale.

● *A set of statements or items are assembled* and from these a number are selected for the final scale. Each item should be chosen to differentiate between those with favourable and those with unfavourable attitudes on the subject under study, because items to which most respondents provide the same answers are unsatisfactory. In attitude scales factual items need to be avoided because respondents may answer according to their knowledge rather than their opinions. Respondents may know, for example, which is the most popular car and their opinions may, therefore, need to be tested by a careful selection of scales.

- *The items included in the final scale may be chosen* either on the basis of an exploratory study in which groups of people are asked to respond to all the items (which is the method used in Likert, Guttman and semantic differential scales – see below); or on the basis of the assessment of a group or judges (which is the method used in Thurstone scales – see below).

Types of scales

Rating scales

A number of problems can arise in the construction of scales, so that a range of methods have been developed in an attempt to solve them.

The simplest method of measuring the strength of an individual's attitude may be to ask the individuals to rate their own strength on a scale. Respondents can be presented with a number of attitude statements and asked which statement comes closest to their own attitude.

RATING SCALES

Respondents may be asked to choose between one of the following statements.

I greatly enjoy mending my own car.
I enjoy mending my own car.
I do not enjoy mending my own car.
I hate mending my own car.

Problems can arise, however, in the construction of a rating scale because if there are too many points on the scale the respondents may be unable to place themselves on it, and if there are too few points the scale may not differentiate adequately between variations in attitudes. The 'error of central tendency' can arise in scales, where respondents avoid placing themselves at either extreme. In the same way, errors of leniency and severity may arise either because respondents dislike being critical or appearing to be critical, or because respondents may set very high standards. There is also the possibility of the 'halo effect' arising so that respondents score favourably on all scales because of a general liking for a commodity or attitude rather than scoring after careful consideration of the scale's meaning.

Thurstone scales

Thurstone scales are attitude scales which use a method of 'equal-appearing intervals' which attempt to form an interval scale of measurement. The procedure in developing a Thurstone scale involves the following.

1   The pooling of all the items that could be included.
2   Selecting a large number which consist of statements on the survey subject

(for example, attitudes to management), ranging from one extreme of favourableness to the other.

3 Reducing this selection by cutting out ambiguous items and duplicate items.
4 Assessment of the items by a large number (around 50) of 'judges' who are asked to assess the items independently. Each judge is given the task of sorting the items into a set number of piles based on an assessment of their degrees of favourableness on the attitude in question and about equally spaced along the attitude continuum (typically seven, nine or eleven piles).
5 Scoring the piles from one to seven (or nine or eleven). For each item a median value is calculated, so that half the remaining judges give the item a lower position and half a higher position on the scale. The interquartile range is calculated also, to measure the scatter of judgements to indicate the extent to which different judges place the items at different parts of the scale.
6 Reducing the list of items further by discarding those with a high scatter because they are in some sense ambiguous or irrelevant and selecting from the remainder around twenty that cover the entire range of attitudes and which appear (judged by the medians) to be about equally spaced along the scale.
7 Embodying the items that have been selected into a questionnaire in random order. Respondents are asked to endorse all the items with which they agree and the average (mean or median) of the median values of all the items endorsed is the respondents' scale score.

Thurstone scales are sometimes termed 'differential' scales because the respondents will agree only with items around their scale position, while disagreeing with more extreme items on either side. This is as distinct from a 'cumulative' scale in which respondents would be expected to agree with all items less extreme than their position and disagree with all those which are more extreme.

Likert scales

Likert scales are attitude scales in which respondents are not asked to decide whether they agree or disagree with an item, but are asked to choose between several response categories, indicating various strengths of agreement and disagreement. The categories are assigned scores and respondents' attitudes are measured by their total scores, which are the sum of the scores of the categories they have endorsed for each of the items. These scales can be described as 'summated rating' scales.

A Likert scale is not an interval scale and conclusions cannot be drawn, therefore, about the meaning of distances between scale positions. It is a reasonable ordinal scale and it is simpler to construct than a Thurstone scale. Although in Thurstone scaling there is an attempt to attain interval measurement, there are doubts as to whether it produces a true interval scale. The objective of a Likert scale is to arrive at the average opinion of people who have contradictory or ambivalent attitudes on a topic and who would not necessarily agree with all the statements below a certain point on a Thurstone scale or disagree with all the statements above it.

Guttman scales

Guttman scales are intricate combinations of the Thurstone and Likert approaches to scaling. **These are attitude scales which start by defining the total attitude to be scaled.** A sample of items representing this total attitude is selected for possible inclusion to be scaled. The scale can be described as 'cumulative', so that respondents are assumed to agree with all the items less extreme than the most extreme one with which they agree.

GUTTMAN SCALES

If mathematical problems are ordered in terms of difficulty, it is reasonable to assume that everyone who obtains a correct answer to the last problem will be able to solve the earlier ones:

$$1 + 2 \; =$$
$$12 + 15 \; =$$
$$25 + 42 - 6 \; =$$
$$(6 \times 4) - 10 \; =$$
$$(15 \times 2) \div (12 - 10 + 6) \; =$$

A major concern in Guttman scaling is the property of reproducibility, which is the ability to reproduce an individual's answers to each item from a knowledge of his total score. Perfect reproducibility would exist with a Likert scale if the correlations between each of the item scores and the total scores were perfect (that is, plus 1), but in practice the correlations fall well short of this value. Likert scales are often found to be poor on reproducibility, as are Thurstone scales although generally to a lesser extent. Guttman scales attempt to provide this quality.

Semantic differential scales

Like Likert scales, **semantic differential scales are summated rating scales.** With Likert scaling there is a range of statements but typically only one standard form of response (strongly agree, agree, etc.), while with the semantic differential there is a range of areas of response but only one issue to evaluate.

With the semantic differential, five-, seven- or nine-point scales are constructed using words (usually adjectives) **and their opposites** (such as good/bad, kind/cruel, true/false, strong/weak, hard/soft) and respondents are asked to indicate their rating of the concept's position on the scale, as in the example below.

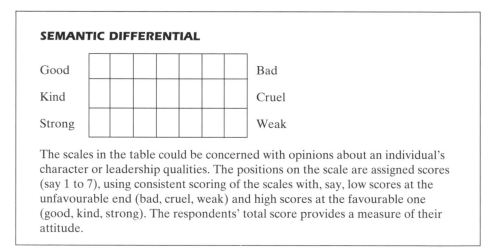

SEMANTIC DIFFERENTIAL

Good								Bad
Kind								Cruel
Strong								Weak

The scales in the table could be concerned with opinions about an individual's character or leadership qualities. The positions on the scale are assigned scores (say 1 to 7), using consistent scoring of the scales with, say, low scores at the unfavourable end (bad, cruel, weak) and high scores at the favourable one (good, kind, strong). The respondents' total score provides a measure of their attitude.

The use of scales

Scales can quantify descriptions and make the assessment of phenomena more exact than it would be otherwise. They can help in the comparison of social phenomena under a variety of conditions in order to ascertain cause and effect. A common application is to provide profiles of objects such as, for example, various competitive brands of goods in a market research inquiry. A scale may help to discover why some brands are more popular than others.

SCALES

A small change in an attribute of a phenomenon, such as the popularity of a particular commodity, may be more rapidly apparent if it is measured on a scale. At the same time, scaling and measurement may increase the possibility of testing research findings through replication. A scale is reliable to the extent that repeat measurements made by it under constant conditions will give the same results, assuming that there is no change in the basic characteristic, such as attitude, which is being measured.

A major difficulty is a scale's 'validity', that is, being sure that the scale is measuring what it is supposed to be measuring. Is the scale, for example, measuring attitudes to the purchase of a particular commodity, or attitudes to the purchase of goods in general? There is usually some reliance on 'face validity', so that the measuring scale seems to make sense on the face of it, in the same way, for example, that the use of occupation makes sense for social class scales (see Table 12.1). Checks can be made as well to see if respondents' reactions to related phenomena are consistent with their scale ratings.

The results of scales need to be interpreted with great care because of the whole problem of 'measuring' and 'quantifying' behaviour. Regional, class and many other factors may influence the results, in that the creation of measuring instruments for social class, for example, may be influenced by the subjective perspectives of the investigator, the respondent and the 'judges'. Scales can be used, however, in a whole range of areas to collect data in market research, social research and psychological testing. They are used on their own or as part of a questionnaire in an overall research design as a quantitative element in the collection of information.

12.8 Financial statistics

Certain aspects of mathematics have a very direct connection with finance; an important example of this is the link between the arithmetic and geometric progressions and simple and compound interest.

Simple interest

An arithmetic progression is where in a series of numbers the difference between them is the same.

ARITHMETIC PROGRESSION

3, 9, 12, 15, 18, 21 (the difference is 3)
$1\frac{1}{2}$, 3, $4\frac{1}{2}$, 6, $7\frac{1}{2}$ (the difference is $1\frac{1}{2}$)

This concept is used for calculations of simple interest.

SIMPLE INTEREST

If £100 is invested for four years at a simple interest rate of 5% per year, at the end of four years the total amount accumulated would be:

£100 + £5 + £5 + £5 + £5 = £120

The total amount has grown in arithmetic progression: £100, £105, £110, £115, £120.

Simple interest can be shown by a formula:

$$A = P(1 + tr)$$

where A is the total amount accumulated
        P is the original investment
        t is the time in years
        r is the rate of interest (given as a decimal or a fraction)

Therefore:

$$A = £100(1 + 4 \times \tfrac{5}{100}) = £120$$

Compound interest

A geometric progression is where in a series of numbers the difference between the numbers is found by multiplying the preceding number by a fixed amount (often called the 'common ratio').

GEOMETRIC PROGRESSION

4, 8, 16, 32, 64 (each number is multiplied by 2 to arrive at the following number).

This concept is used in calculations of compound interest. When a sum of money has been invested (in a bank account or building society) the interest payment is often reinvested rather than withdrawn and spent. This is the basis of compound interest.

COMPOUND INTEREST

If £100 is invested for four years at a compound interest rate of 5% per annum, at the end of four years the total amount accumulated would be:

original investment = £100
end of year 1   = £105
end of year 2   = £110.25 (£105 + $\tfrac{5}{100}$ × £105)
end of year 3   = £115.76
end of year 4   = £121.55

Compound interest can be shown by a formula:

$$A = P(1 + r)^t$$

where A is the total amount accumulated
       P is the original investment
       r is the rate of interest
       $(1 + r)$ is the 'common ratio'
       t is the time in years

The original investment is multiplied by the common ratio to the power of t, that is, the number of years which the investment lasts.

Therefore:

$$A = £100(1 + \tfrac{5}{100})^4 = £121.55$$

Notice that P is known as the 'present value' of A at a compound interest rate (r) in t years from now. So that £100 is the present value of £121.60 at a 5% rate of interest in four years time.

Money invested at compound interest quickly builds up to a large sum: £100 at 5% over 10 years would produce an accumulation of £162.89 while at simple interest this would be £150. This is why compound interest has become so important. Most savings such as life assurance policies and pension funds are partly built up from the compound interest they earn.

Present values

The concept of present value works in the opposite direction to the calculation of compound interest. The kind of question asked is: 'What sum of money, if invested at an interest rate of 10% per annum compounded annually, will give £100 in five years time?'

The formula is:

$$P = \frac{A}{(1 + r)^t}$$

where P is the original investment or the present value

A is the total amount accumulated

r is the rate of interest

t is the time in years

Therefore (in answer to the question posed above):

$$P = \frac{100}{(1 + \tfrac{10}{100})^5}$$

$$= \textbf{£62}$$

Thus £62 is the present value (or discounted value) of £100 due at the end of five years: £62 now is equivalent to £100 in five years time. This has nothing to do with inflation. It can be argued that £1 now is always worth more than £1 in the future, simple because the £1 can be used to earn more money. When a business borrows money, it is in effect exchanging a larger sum of future money for a smaller sum of present money, which it can use profitably.

When calculating present values, the discount rate used is normally the pre-vailing rate of interest on money borrowed.

Present values have a number of financial applications. One example is **discounted cash flow which involves the calculation of the present value of a series of future cash flows. Cash flow is the difference between the money flowing into a business** (from receipts) **over a certain period and the money flowing out** (from payments).

By calculating present values or the net present value of a proposed invest-ment project a business can ascertain its present worth. Faced with a choice between alternative investments, businesses can calculate their net present values to provide a valid means of comparison.

NET PRESENT VALUES

A firm has a choice between two plants, one of which costs £50 000, the other £40 000. Each has a useful life of four years. Which plant should the firm purchase if the discount rate is 20%?

	Cost	Estimated annual cash flow (£)			
Plant A	−£50 000	+10 000	+20 000	+40 000	+30 000
Plant B	−£40 000	+10 000	+20 000	+15 000	+5000

Plant A's net present value (in £000s):

$$-50 + \frac{10}{1.2} + \frac{20}{1.2^2} + \frac{40}{1.2^3} + \frac{30}{1.2^4} = 9.84 \text{ (in £000s)} = £9840$$

Plant B's net present value (in £000s):

$$-40 + \frac{10}{1.2} + \frac{20}{1.2^2} + \frac{15}{1.2^3} + \frac{5}{1.2^4} = -6.69 \text{ (in £000s)} = -£6690$$

The cash flow for each year is divided by 1 plus the rate of interest (20% or 0.2) to the power of the time in years (as in the formula for present values above). Plant A has the higher net present value. The negative net present value of plant B shows that this plant will yield a return on investment which is less than the current discount rate. A project with a negative net present value is not viable since either the firm will not be able to cover the cost of borrowing or, if it has spare cash, it can invest it more profitably elsewhere.

Discounted rate of return

This is concerned with the discount rate which will give a net present value of zero.

In the previous example, if the firm has to buy plant B or give up the project, the problem is to find the discount rate that will give a zero net present value and therefore make the project viable. The 20% discount rate applied in the example gives a negative net present value. For a zero net present value, the discount rate must be less than 20%. To determine what this is it is necessary to proceed by trial and error.

A discount rate of 10% gives a net present value of +£0.3200. This is just positive. A discount rate of 11% gives a net present value of −£0.500. This is just negative. Therefore the expected returns from this project will repay the original investment of £40 000 if the firm can borrow the money at an interest rate of 10% or less.

Decisions

Techniques such as those discussed above may help a firm to make a decision on investment, loans and borrowing. In fact these decisions are complicated by taxation, government grants and so on. These can be appropriately discounted.

Many investment decisions are made on the basis of other factors, other costs and benefits which may include such factors as the environment, industrial relations, 'hunches' and so on which are not easily 'discounted'.

Spreadsheets

A spreadsheet is a matrix on which calculations can be made. Electronic (computerised) spreadsheets are a convenient way of setting up a range of charts, records and tables, including profit and loss accounts, budgeting charts and sales forecasts.

12.9 Information technology

Information technology (IT) is the acquisition, production, transformation, storage and transmission of data by electronic means. It is concerned with access to and the processing of knowledge and information. It uses vocal, pictorial, textual and numerical information to facilitate the interaction between people, and between people and machines, typically through a microelectronics-based combination of computing and telecommunications interacting with other technologies. It includes the social, economic and cultural implications of these processes and their applications. Information technology has greatly increased the ability to manipulate numbers and it has emphasised the importance of being numerate as well as literate (see Section 1.3).

INFORMATION TECHNOLOGY

IT includes: control technology, systems management, information engineering and the manipulation of data of all kinds. It is based on key enabling technologies such as Software Engineering, Very Large Scale Integration (VLSI), Man Machine Interfaces (MMI) and Intelligent Knowledge Based Systems (IKBS). The so called Fifth Generation of computers have been developed to include artificial intelligence and expert systems. The Fifth Generation is the next step of development following on from the First Generation (approximately 1940–52) with vacuum tubes as the electronic component, the Second Generation (1952–62) with transistor electronics, the Third Generation (1964–71) which used integrated circuits and the Fourth Generation (1971–1990s) with large-scale integrated circuits. The research into

---

artificial intelligence is an attempt to produce a computer that thinks. The 'Turing Test' of an intelligent machine or program is whether it can perform a task, such as answering questions, with a result that is indistinguishable from that of a human being. Expert systems are based on people who can negotiate an agreed interpretation of a particular subject with the help of special knowledge or opinions.

The manufacturing economy is being succeeded by the 'information, knowledge or post-industrial society', the characteristic of which is that the majority of the population becomes involved in information handling for its living, and correspondingly the proportion of the population directly involved in manufacturing processes will be a small minority. The ability to work with figures as well as words has increased in importance as a result of these developments. Because of this, knowledge is becoming the new wealth of nations, and information, the raw material of IT, is assuming the character of the fourth or fifth 'resource' or 'factor of production' along with land, labour and capital (and enterprise). The implications of these developments include both the speed of change and the intensity of change (such as the effects of the introduction of plastic money on currency handling). A range of computer databases are becoming the focus for a variety of knowledge-based activities and the IT economy makes it necessary for everybody to have access to data and to the control of systems through such facilities as programming, modelling and simulations.

The software (a set of programs) brings the hardware of computers to life; it is 'soft' because it can be easily changed whereas the hardware remains the same, at least for some time. Each program consists of a sequence of instructions which can be manipulated by a computer and the word 'program' is sometimes used synonymously with the word 'algorithm'. **An algorithm is the logical idea behind the program, it is a process of rules for calculation, a strategy for solving a problem.** For example, the statement 'explain something useful about the numbers 4 and 6' is not an algorithm, while the statement 'add 4 and 6, divide the sum by 2 and report the answer' is an algorithm because it can be translated into a suitable programming language and processed through a computer.

MODELS

Simulation and model building are important applications of computer technology in a range of areas. **Models are theoretical constructions in which interrelationships are expressed in diagrammatic, verbal or mathematical terms.** A model or simulation can be a large- or small-scale replica or imitation of the thing or system being studied, such as a model ship or a simulation of a court case. In the social sciences, models are an attempt to imitate a real-life situation. In aspects of economics and the control of the economy, for example, simulations and models have been developed based on statistical data and relying on the manipulation of this data. A simulation which has the objective of maximising welfare over a period of years, for example, will include decisions to be made on the level of government expenditure, the money supply and tax rates in order to eliminate inflationary or deflationary trends. Computer models have been created for setting the task of achieving a steady rate of growth plus equilibrium in the balance of payments, by using foreign exchange reserves, interest rates and the budget as policy instruments. These macroeconomic models have been complemented by microeconomic simulations in relation to such topics as price and output under particular market conditions (such as oligopoly), and the marketing mix with reference to price, advertising and sales. There are suites of statistical programs for manipulating the data and for testing hypotheses.

Information technology increases the amount of information that is available, increases the facility of storing it and the speed of retrieving it, while computers have given a new significance to numbers because the computer allows the quantification of all kinds of data. IT enables statistical models and simulations to be developed and manipulated with great ease so that a range of theories and hypotheses can be tested. All of these attributes emphasise the need identified at the beginning of this book for people to be able to absorb, select and reject information in order to pick out what is interesting and useful. This highlights the use of statistics as a scientific method for collecting, organising, summarising, presenting and analysing data.

12.10 The next step

This chapter outlines areas of further study in statistics with examples of the use of statistics in these areas. The application of statistical techniques to a variety of topics and to aspects of work and research is open-ended and has been given increasing impetus by developments in information technology. The quantity of data has increased immensely and statistical techniques can play an important role in helping to ensure an improvement in quality. The processes of model building, experimentation and testing involve scientific methods and a logical approach which encourage a rigorous use of data.

This book provides the basic framework for understanding and interpreting statistics; at the same time applications are illustrated and the way forward is suggested. The next step is to travel further along the statistical road in the direction of both the interpretation of numerical information and its application.

ASSIGNMENTS

1   Explain the meaning of the following equation:

$$\sum_{i=3}^{8} x_i = x_3 + x_4 + x_5 + x_6 + x_7 + x_8$$

2   Construct and carry out an experiment on a small group of people as a pilot study to prepare for a fully controlled investigation.

3   Discuss the difficulties of carrying out experiments in the social sciences and whether they are so lacking in scientific method as to be useless.

4   Consider the various uses of statistical tests.

5   Draw up an attitude scale, test it on a small number of people and interpret the results.

6   What sum of money, if invested at an interest rate of 20% per year compounded annually, will give £500 in three years' time?

7   Find examples of the use of compound interest in the economy. Write a report about its use.

Appendixes

Further reading

Practice and revision

Tim Hannagan, *Work Out Statistics* (Macmillan, 1995). This is in the Macmillan 'Work Out' series and is designed specifically for GCSE level statistics. It includes large numbers of worked examples, with working and answers; and questions and exercises with solutions. These are particularly useful, in fact essential, for understanding and practising calculations. There are, for instance, 48 worked examples of averages and 26 questions and exercises with answers.

More advanced statistics

M. R. Spiegal, *Theory and Problems of Statistics* (McGraw Hill, 1960). This remains the most comprehensive, advanced text on statistical techniques.
C. A. Moser and K. Kalton, *Survey Methods in Social Investigation* (Heinemann, 1971). This is the classic book on survey methods.

Examples of the use of statistics

Social Trends, Central Statistical Office (HMSO, annually). This provides as many tables of figures, graphs and diagrams as most people would want, but there are many other government and international publications full of statistical data. Company annual reports also provide many examples of the use of statistics.

A2 Area table

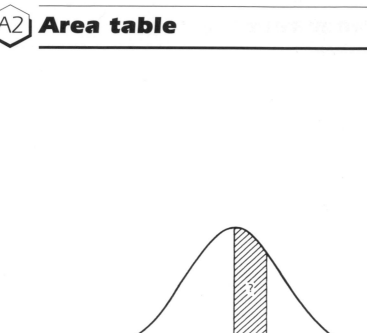

Areas under the normal curve

x is the distance from the mean measured in standard deviations to arrive at z values:

$$z = \frac{x - \bar{x}}{\sigma}$$

z	Area	z	Area
0.0	0.0000	1.5	0.4332
0.1	0.0398	1.6	0.4452
0.2	0.0793	1.7	0.4554
0.3	0.1179	1.8	0.4641
0.4	0.1554	1.9	0.4713
0.5	0.1915	2.0	0.4772
0.6	0.2257	2.1	0.4821
0.7	0.2580	2.2	0.4861
0.8	0.2881	2.3	0.4893
0.9	0.3159	2.4	0.4918
1.0	0.3413	2.5	0.4938
1.1	0.3643	2.6	0.4953
1.2	0.3849	2.7	0.4965
1.3	0.4032	2.8	0.4974
1.4	0.4192	2.9	0.4981
		3.0	0.4987

Basic mathematics

A3.1 The need to use mathematics

Both the descriptive and inferential aspects of statistics require the use of basic mathematics. Once information has been collected by a full survey or a sample, it has to be put into a form which can be used easily. The raw survey data has to be processed, analysed, summarised and presented. Much of this process is descriptive in the form of reports, tables, graphs and averages but also inferences can be drawn from the data based on its accuracy.

A3.2 The vocabulary of mathematics

It can be argued that the main purpose of a language is to enable people to exchange ideas with the minimum of effort and the maximum of clarity. Mathematics has its own language, with a vocabulary of its own in the form of numbers and symbols (see Section A3.16). **The basic vocabulary of mathematics is a number system.**

The decimal system

This is based on groups of 10 (from the Latin *decem*, meaning ten). The system was invented by the Hindus about 1500 years ago and was passed on by the Arabs towards the end of the eleventh century.

The system is based on the idea of positional value and a base of 10. All numerals are constructed using 10 basic symbols, and in writing numerals the position of any given numeral is significant.

The basic symbols are 0 to 9. To represent numbers ten times as large, these digits are shifted one position to the left, the digit 0 being used as a position indicator: 10, 20, 30, . . ., 90. Further increases by a factor of ten are indicated by further shifts in position: 100, 200, 300, . . .

A decrease by a factor of ten is shown by shifting the digits one position to the right, the digit 0 again being used when necessary to indicate position: one-tenth is 0.1, one-hundredth is 0.01, one-thousandth is 0.001.

THE DECIMAL SYSTEM

Thousands	Hundreds	Tens	Units
1	4	6	2
5	9	4	1
	7	3	2
2	0	3	6
10	1	7	1

The positions of the numbers in the columns are important; a number in any column represents ten times the same number in the column on its right and one-tenth of the same number in the column on its left.

The binary system

This has a base of 2 (and is used in computing). **The two digits used are 0 and 1 and any number can be represented by locating these digits in appropriate positions.** Zero is 0, one is 1, two is 10, three is 11, four is 100, five is 101, six is 110, seven is 111, eight is 1000.

An increase in a number by a factor of 2 (that is, doubling it) is shown by shifting the digits one position to the left, 0 being used to indicate position. A decrease by a factor of 2 (halving the number) is shown by shifting the digits one position to the right.

Systems based on numbers other than ten or two are possible, but these two systems are the most common.

A3.3   Basic arithmetic

The four basic arithmetical operations are addition, subtraction, multiplication and division, with the symbols +, −, ×, ÷ respectively.

Addition is the process of putting numbers together. The sign + (or *plus* from the Latin for 'more') means that the number following it is to be added to the number preceding it. The answer obtained by adding numbers together is called the 'sum' (with the symbol Σ or sigma), and the answer is indicated by the sign = (equals).

Positive numbers are normally written without a plus sign, while with negative numbers the minus is always written. The value of a number without its sign is called its 'absolute' value. The number 2 is the absolute value of both -2 and $+2$.

Subtraction is the process of finding the difference between two numbers. It is the inverse of addition. The sign $-$ (or *minus*, from the Latin for 'less') indicates that the number following is to be taken away from the number preceding it.

Multiplication is the process of finding the sum of a number of quantities which are all equal to one another. It is essentially a short-cut version of adding when all the numbers are the same. Therefore 8×6 means $8 + 8 + 8 + 8 + 8 + 8 = 48$. Multiplication is indicated by the sign $\times$ (or *times*) and the result of the multiplication is called the 'product'.

When two negative numbers are multiplied together they make a positive product, while a positive and negative number multiplied produces a negative product ($-8 \times -6 = 48$, but $8 \times -6 = -48$).

Division is the process of finding out how many times one number is contained in another number. It is the inverse of multiplication. Division is indicated by the sign $\div$ (or *divided by*).

A3.4 The sequence of operations

Addition, subtraction, multiplication and division must be carried out in the correct order in mathematical calculations. The mnemonic *BODMAS* summarises this sequence:

*B*rackets
*O*f
*D*ivision
*M*ultiplication
*A*ddition
*S*ubtraction

This could be pronounced *B—ODM—AS*, because *ODM* have equal priority and *AS* have equal priority. 'Of' stands for multiplication, as in '$\frac{1}{2}$ of 10'.

SEQUENCE OF OPERATIONS

1 $5 + 2 \times 3 = 11$ *not* 21 ($2 \times 3 \rightarrow 6 + 5 \rightarrow 11$, *not* $5 + 2 \rightarrow 7 \times 3 \rightarrow 21$)
2 $(5 + 2) \times 3 = 7 \times 3 = 21$

These examples illustrate the fact that:

- brackets should be evaluated first;
- if an expression contains only pluses and minuses or only multiplication and division, then the sum should be calculated by working from the left to the right;
- multiplication and division should be calculated before addition and subtraction.

A3.5 Simple arithmetic

The following sums are included to illustrate the rules of simple arithmetic. If you have any doubts about these rules, then work the sums out before looking at the answers and explanations.

SIMPLE ARITHMETIC

1   $30 - 9 + 3$ In this case the working is from left to right (24).
2   $8 \times 3 - 2$ The multiplication comes before the subtraction (22).
3   -5×3 A minus times a plus equals a minus (-15).
4   -5×-3 Two minuses make a plus, in the same way that a double negative cancels out: '*not im*possible' means that it is possible (15).
5   $4 + 7 \times 3$ Multiplication comes before addition (25).
6   $4(3)$ A bracket indicates multiplication. This is another way of writing 4×3 (12).
7   $4(3 + 1)$ The bracket always comes first (16).
8   $(2 - 1)(7 + 5)$ The brackets come first, the two brackets indicate multiplication (12).
9   $5 - 3 \times 4 + 5$ Multiplication comes before subtraction and addition (-2).
10   $122 \times 28 \div 7 + 60$ Multiplication and division come first (but have equal priority), addition last (548).

A3.6 Fractions

Fractions are units of measurement expressed as one whole number divided by another. They allow the consideration of units of measurement smaller than a whole number. The term 'common fraction' is used to emphasise the distinction from the decimal fraction: $\frac{1}{2}$ as opposed to 0.5.

$$\text{Common fraction} = \frac{\text{Numerator}}{\text{Denominator}}$$

A 'proper fraction' is one where the numerator is less than the denominator ($\frac{1}{2}$ or $\frac{3}{4}$). An 'improper fraction' is where the numerator is greater than the denominator: $\frac{53}{10}$. This can be reduced to a whole number and a proper fraction: $5\frac{3}{10}$.

The addition and subtraction of fractions

Fractions often occur when things are measured or when they are divided. The classic example is dividing a cake into slices. Assume that the cake is divided equally between three people; it will be divided into three equal slices or thirds:

$$\frac{1 \text{ whole cake}}{3} = 3 \text{ slices, each } \tfrac{1}{3} \text{ of a whole cake.}$$

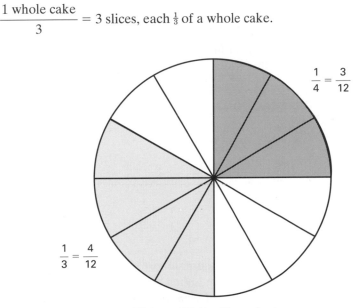

$$\frac{1}{4} = \frac{3}{12}$$

$$\frac{1}{3} = \frac{4}{12}$$

Figure A3.1   Slices of cake

If one person eats $\frac{1}{3}$ of the cake and then $\frac{1}{4}$ of the cake, in total that person has eaten $\frac{1}{3} + \frac{1}{4}$ pieces of cake. To add these the common denominator is found, that is, the number of slices into which both $\frac{1}{3}$ and $\frac{1}{4}$ can be divided with whole numbers. If a cake is divided into twelve slices, it is possible to count how many slices of cake the person has consumed; $\frac{1}{3} = 4$ slices and $\frac{1}{4} = 3$ slices. Therefore $\frac{1}{4} + \frac{1}{3} = \frac{3}{12} + \frac{4}{12} = \frac{3+4}{12} = \frac{7}{12}$.

If the person was told that $\frac{5}{12}$ of cake would be taken away from $\frac{7}{12}$ of cake to give to other people, then that person would have $\frac{7}{12} - \frac{5}{12} = \frac{2}{12}$ or $\frac{1}{6}$th of the cake left. This can be checked in Figure A3.1.

The multiplication and division of fractions

To multiply fractions the numerators are multiplied together to obtain the numerator of the answer, and the denominators are multiplied together to obtain the denominator of the answer.

> **MULTIPLICATION OF FRACTIONS**
>
> $\frac{2}{3} \times \frac{4}{12} = \frac{8}{36} = \frac{2}{9}$

To divide by a fraction, the fraction is multiplied by its inverse.

> **DIVISION OF FRACTIONS**
>
> $\frac{4}{7} \div \frac{5}{8} = \frac{4}{7} \times \frac{8}{5} = \frac{32}{35}$
>
> $125 \div \frac{1}{5} = 125 \times \frac{5}{1} = 625$

To divide mixed fractions they must first be made improper (a 'mixed fraction' is one that contains whole numbers as well as fractions).

> **DIVISION OF MIXED FRACTIONS**
>
> $3\frac{2}{5} = 3 + \frac{2}{5} = \frac{15}{5} + \frac{2}{5} = \frac{17}{5}$
>
> $\therefore \quad 3\frac{2}{5} \div 2\frac{4}{7} = \frac{17}{5} \div \frac{18}{7} = \frac{17}{5} \times \frac{7}{18} = \frac{119}{90} = 1\frac{29}{90}$

A3.7 Decimals

A decimal number is a fraction whose denominator is 10, or 100, or 1000, or any power of ten. For instance, the decimal number 'three-tenths' is written $\frac{3}{10}$ or 0.3. The decimal point divides the whole number from the fraction. 1.3 equals the whole number 1, plus the fraction 0.3. The decimal point should not be confused with a full stop. The decimal point can be written above the line, as in 0·3, to avoid this ambiguity.

When adding and subtracting numbers which include decimals, the decimal points must be kept underneath one another to avoid the difficulties of place value.

> **DECIMALS**
>
> Add 1.16, 2.75, 0.08, 8.057.
>
> ```
>   1.16
>   2.75
>   0.08
>   8.057
> ─────────
>  12.047
> ```

Applications of the decimal system include money and the metric system.

Decimal currency

The decimal currency of the UK is based on the pound sterling (£) and is divided into 100 pence (p), just as the US dollar is divided into 100 cents. Sums of less than £1 ($1) in value can be written in two ways:

- 55p (55 cents)
- £0.55 ($0.55)

If the number of pence is less than 10, there should be a 0 in the ten pence column to indicate that there are not any 10 pence pieces involved:

£0.08 = 8p   £0.80 = 80p

The metric system

There is a general world trend towards using the metric system of weights and measures. **The name is derived from the basic unit of length, the metre.**

The International System of Units (Système Internationale, referred to as SI) covers all types of measurement, for example:

the metre (m) for length
the litre (l) for volume and capacity
the kilogram (kg) for weight
the degree Centigrade (°C) for temperature
the second (s) for time.

The tables for all these measures are based on units of ten:

THE METRIC SYSTEM

Table of length

10 millimetres (mm)	= 1 centimetre (cm)
100 centimetres (cm)	= 1 metre (m)
1000 metres (m)	= 1 kilometre (km)

Table of capacity

1000 millilitres (ml)	= 1 litre (l)
1000 litres (l)	= 1 kilolitre (kl)

Table of weights

100 milligrams (mg)	= 1 gram (g)
1000 grams (g)	= 1 kilogram (kg)
1000 kilograms (kg)	= 1 tonne or metric tonne (t)

The imperial system

The metric system has generally replaced the imperial system of weights and measures, but the imperial system is still used and is likely to be of both practical and historical interest for some time. The imperial system is based on:

the foot (ft) for length
the pint (pt) for volume and capacity
the pound (lb) for weight
the degree Fahrenheit (°F) for temperature
the second (s) or minute (min) for time.

The tables for all of these are based on a variety of units of numbers;

THE IMPERIAL SYSTEM

Table of length

12 inches	= 1 foot
3 feet	= 1 yard
1760 yards	= 1 mile

(1 metre = 39.37 inches or 3 feet 3.37 inches)

Table of capacity

2 pints	= 1 quart
4 quarts or 8 pints	= 1 gallon

(1 litre = 1.76 pints)

Table of weight

16 ounces (oz)	= 1 pound (lb)
14 pounds (lb)	= 1 stone
112 pounds (lb)	= 1 hundredweight (cwt)
20 hundredweight (cwt)	= 1 ton

(1 kilogram = 2.2 pounds (lb))

Temperature

The freezing point of water in Celsius is at $0\,°C$, in Fahrenheit at $32\,°F$. Conversion formulas are:

$$C = \tfrac{5}{9}(F - 32) \text{ and } F = \tfrac{9}{5}C + 32$$

Therefore if $F = 60\,°F$ then:

$$C = \tfrac{5}{9}(60 - 32)$$
$$= 15.55\,°C$$

If $C = 15\,°C$ then:

$$F = \tfrac{9}{5} \times 15 + 32$$
$$= 59\,°F$$

A3.8 Percentages

'Per cent' (or percent in the USA) means per hundred (from *centum*, the Latin for hundred). Therefore 50 per cent is 50 per hundred or 50 out of a hundred, that is, one half. The sign for a percentage is % which contains the one, zero, zero of 100.

USING PERCENTAGES

- To change a fraction to a percentage multiply by 100: $\frac{1}{4}$ equals 25% $(25 = \frac{1}{4} \times 100)$.
- To change a decimal to a percentage multiply the decimal by 100 and put the % after the result: $0.55 \times 100 = 55\%$.
- To change a percentage to a fraction, divide by 100: $50\% = \frac{50}{100} = \frac{1}{2}$.
- To change a percentage to a decimal, divide by 100 by moving the decimal point two places to the left: $13.96\% = 0.1396$.
- Finding a percentage can be seen by example:

  £7 as a percentage of £50 $= \frac{7}{50} \times 100 = \frac{700}{50} = 14\%$

  5% of £4.50 $= \frac{5}{100} \times £4.50 = £\frac{22.50}{100} = 22.5$p

---

A3.9   Powers and roots

Multiplication is a shorthand way of representing a series of additions; **'powers' are a shorthand method of representing a series of multiplications.**

$4 \times 4 \times 4$ is four raised to the power of three, or $4^3 = 64$

'Four to the power two' (4^2) is 'four squared'.

The square root of a number is that number which when multiplied by itself gives the original number. Square root is shown by the sign $\sqrt{}$. The square roots of many numbers are difficult to calculate exactly except with a calculator or computer, although they can be estimated through trial and error.

SQUARE ROOTS

The square root of 16 is 4, because $4 \times 4 = 16$.
    Therefore $\sqrt{16} = 4$
    The square root of 30 lies between 5 and 6 ($5 \times 5 = 25$, $6 \times 6 = 36$). A close estimate would be 5.5. Correct to two decimal places the square root of 30 is 5.48 ($5.48 \times 5.48 = 30.02$).

A3.10   Ratios

A ratio is a relationship between two quantities expressed in a number of units which enables comparison to be made between them.

RATIOS

1 Two motor-cars may be travelling at different speeds, say 60 km per hour and 30 km per hour. The ratio of speeds to one another is said to be: 60 : 30, or 6 : 3, or 2 : 1.

2 Three-quarters of the annual output of a factory may consist of commodity A and one-quarter of commodity B. The ratio of output can be said to be 3 : 1. For every 3 units of A produced in a year, 1 unit of B is produced.

A3.11 Proportions

Many business calculations are based on simple proportions.

PROPORTIONS

1 If 5 kilograms of copper cost £24 what will 12 kilograms of copper cost? If the price is the same for every kilogram, then the higher cost of the 12 kilograms will be in direct proportion:

Cost of 5 kg = £24
Cost of 1 kg = £4.8 (£24 ÷ 5)
Cost of 12 kg = £57.6 (£4.8 × 12)

2 If it costs £112 to stay at a hotel for 7 days, how much will it cost for 15 days at the same rate?

Cost of 7 days = £112
Cost of 1 day   = £16 (£112 ÷ 7)
Cost of 15 days = £240 (£16 × 15)

A3.12 Elementary algebra

Algebra provides a method of abbreviating information without loss of clarity or accuracy:

ALGEBRA

The contents of a basket of fruit could be written in an abbreviated form as: 10 As, 5 Os, 8 Bs, where A = apples, O = oranges and B = bananas. Equally these three types of fruit could be labelled x, y and z. If another basket of fruit was said to contain $15x$ and $10y$, this could be translated as 15 apples and 10 oranges.

The expression $15x$ means 15 times x. Therefore, if x is also a number, say 3, then the value of $15x$ when $x = 3$ is 45.

This should not be confused with the use of letters such as x and y as symbols in statistical calculations, where they often denote two variables.

Simple equations

$x + 6 = 8$

This means that some number plus 6 equals 8. Since 2 is the only number which when added to 6 equals 8, then x must equal 2.

If $x + 6 = 8$
then $x = 2$

$x + x + x$ is written $3x$ and means 3 times x.

If $6x = 30$
then $x = 5$

An algebraic formula could be:

$T = a + 2b + 5g$

where a = apples, b = bananas, g = grapes and T is the total number of units.

If $a = 2, b = 5, g = 12$

then $T = 2 + 10 + 60 = 72$

A3.13 Levels of measurement

In mathematics there are different levels of measurement, from a basic labelling system to a system which indicates the value of one number against another.

Nominal scales

Nominal scales are a method of classification and in this the function of numbers is the same as names: to label categories. There is no implication that one is better or greater than another.

NOMINAL SCALES

Hotel rooms have numbers to distinguish them, but usually not to indicate which is the best or worst room. There is no point in adding up hotel room numbers; it may be useful to know that there are ten rooms but it is not likely to be useful to know that the numbers add up to 55.

Nominal scales possess the properties of symmetry and transitivity. Symmetry means that a relationship between *A* and *B* is also true between *B* and *A*. If *A* is opposite to *B*, then *B* is opposite to *A*.

Transitivity means that if $A = B$ and $B = C$, then $A = C$. If *A* is the same age as *B*, and *B* is the same age as *C*, then *A* must be the same age as *C*.

Ordinal scales

This is the ordering of categories with respect to the degree to which they possess particular characteristics, without being able to say exactly how much of the characteristic they possess.

ORDINAL SCALES

Example 1
Workers can be classified into 'unskilled', 'semi-skilled' and 'skilled' without giving the exact interval between these classifications.

Example 2
Three factories may be put into order by size: 1, 2 and 3. This will not give any indication of relative sizes. In fact the factories may employ 10 000 people, 8000 people and 30 people respectively, but these differences are not shown on the ordinal scale.

Ordinal scales are asymmetrical. This means that a special relationship may hold between *A* and *B* which does not hold between *B* and *A*. If *A* is greater than *B*, then *B* cannot be greater than *A*.

Ordinal scales do have the property of transitivity. If *A* is greater than *B* and *B* is greater than *C*, then *A* will be greater than *C*.

Interval scales

These scales not only rank objects with respect to the degree with which they possess a certain characteristic, but also indicate the exact distances between them. This requires a physical unit of measurement which can be agreed upon as a common standard and that can be applied over and over again with the same results.

INTERVAL SCALES

Examples of these units of measurement are the metric system and the imperial system. Length, for example, is measured in metres or feet. There are no such units of intelligence or authoritarianism which can be agreed upon.

Given a unit of measurement, it is possible to say that the difference between scores is 20 units, or that one difference is twice as large as a second. Scores can be added and subtracted and so on.

The scales compared

An ordinal scale possesses all the properties of a nominal scale plus ordinality. An interval scale has all the properties of a nominal and ordinal scale plus a unit of measurement.

The cumulative nature of these scales means that it is always possible to drop back one or more levels of measurement in analysing data. Sometimes this is necessary when statistical techniques are unavailable or unsatisfactory for handling the variable on an interval scale. However, by using a lower scale information is lost:

USE OF SCALES

If it is known that A has an annual income of £15 000, B has an income of £14 000 and C has an income of £5000, it is possible to use the ordinal scale and say that A has the highest income, B the second highest and C the lowest. In order of size of incomes the ranking is $A = 1, B = 2, C = 3$.

However, using the ordinal scale in this example throws away the information that the difference in incomes is small between A and B and much larger between A/B and C, and that A has an income that is three times as great as that of C. Therefore it is advantageous to make use of the highest level of measurement that can be used.

A3.17 Symbols of mathematics

This list includes the main mathematical symbols used in statistics:

x is a collective symbol meaning all the individual values of a variable.

y is an alternative symbol to x. It is used where there are two sets of variables and x has already been used to indicate the first.

$\bar{x}$ (bar x or x bar): a bar over a variable symbol indicates that it represents the arithmetic mean of the values of that variable.

μ (pronounced 'mew'): the population mean.

n stands for the number of items in a collection of figures (10 apples, $n = 10$).

f stands for *frequency*: that is, the number of times a given value occurs in a collection of figures.

Σ means the *sum of* (Σ is the Greek capital S or sigma).

σ (lower case sigma) stands for the *standard deviation*.

d stands for *deviation*, which is the difference between two values.

r stands for the *coefficient of correlation*.

r' stands for the *coefficient of rank correlation*.

$=$ means *equals*.

$\approx$ means *approximately equals*.

$\neq$ means *not equal to*.

$>$ means *greater than* or *larger than* or *more than*.

$<$ means *smaller than* or *less than*.

ASSIGNMENTS

1   Calculate the following:
    (i) $6 + 4 \times 2$
    (ii) $-2(2 - 1)$
    (iii) $\frac{1}{4} - \frac{1}{8}$
    (iv) $6\frac{3}{4} - 3\frac{7}{8}$
    (v) $0.92 - 0.42$
    (vi) $9 \div 0.03$
    (vii) £6 as a percentage of 510
    (viii) 20% of £5
    (ix) 2^7
    (x) $\sqrt{0.01}$
    (xi) $(1.4)^2$
    (xii) $5x + 6y$ when $x = 3$ and $y = 4$
2   In an office of 640 employees $\frac{3}{8}$ are women and $\frac{5}{8}$ are men. How many (a) women and (b) men are there?
3   Discuss why the highest possible level of measurement should be used.

A4 How to pass statistics examinations

Introduction

The main objective of this book is to help candidates pass statistics examinations such as:

GCSE
GNVQ
BTEC National Level/Higher National Level
Institute of Marketing
Institute of Chartered Secretaries and Administrators
Institute of Cost and Management Accountants
Institute of Chartered Accountants
Association of Certified Accountants
Institute of Personnel Management

The book is also an introduction to statistics for the general reader and for a wide variety of people studying for degree and diploma qualifications and GCE 'A' and 'AS' Level examinations. Candidates who work their way through this book should understand numerical concepts and be proficient in basic statistical methods at a level which will enable them to pass the examinations listed above and similar statistics examinations.

It would be a considerable help to study *Work Out Statistics* Tim Hannagan (Macmillan Education, 1995), in particular to concentrate on the worked examples and the questions and exercises, because they provide excellent practice for examinations.

An examination involves answering questions over a few hours on what has been learnt over a year or two and is a test of the ability to extract the central points from the syllabus. However much work has been carried out during the time leading up to an examination, a period of intensive revision will be necessary. Revision should start at least six to eight weeks before the examinations and should be based on a timetable so that each topic is revised in turn with more time concentrated on a candidate's 'weak' areas.

It is useful to have built up a series of notes covering the whole syllabus to provide a basis for revision. The content of these notes should be determined by individual needs for reminders of statistical methods and applications. Headings and sections, lists of formulae and methods all help in 'last minute' revision.

Examinations are based on knowledge and memory as well as understanding. In statistics this means that techniques need to be followed, formulae and definitions remembered, and applications and examples prepared so that they can be used in a relevant way to illustrate and explain concepts.

Above all in preparing for a statistics examination, practice is essential, particularly of calculations and preferably under examination conditions.

Planning for a statistics examination

In planning for a statistics examination it is essential to obtain certain things and to practise calculations, as follows.

Obtain the syllabus for the examination

The syllabus will show what will be examined and how it will be examined. It is important to know the type of questions which will be asked, whether there is a choice and how many questions have to be answered. In most statistics examinations calculators can be used and papers are set assuming their use. This means that questions may include a number of calculations which would take a long time without a calculator but can be completed quickly with one.

Statistics exams seldom involve long essay questions but tend to include a variety of other types of questions, such as:

- *calculation questions*, where a calculation or a number of calculations are asked for;
- *problem-solving questions*, where a problem is posed which might involve in its solution a combination of calculations, analysis, comment and discussion;
- *interpretation questions*, where an analysis is asked for of a table, diagram or graph or of a solution to a calculation;
- *multiple-choice or 'objective' test questions*, where several alternative answers are given and the correct one has to be ticked;
- *short-answer questions*, where a two-line written answer is required to what is often an interpretation question.

There is a growing tendency amongst examination bodies to ask a variety of types of question. Multiple-choice and short-answer questions are asked in order to test the whole syllabus, while longer problem-solving questions test particular parts of the syllabus in more detail.

Obtain the marking schemes and examiners' reports for the examination.

In answering examination papers it is essential to know how marks are distributed over the questions and as far as possible for each part of a question, because this distribution will determine the length of time spent on each section of the paper.

The instructions on the examination paper will invariably indicate the distribution of marks between the questions, sometimes by stating that 'all questions carry equal marks'. In the case of some examinations the mark for each part of a question is shown on the paper; other examining boards include marking schemes in examiners' reports.

In short calculation questions most, if not all, of the marks are for a correct answer; in problem-solving questions marks are given for the various sections of the question and marks may be given for correct methods or consistent answers even when the solution is not entirely accurate; while interpretation questions will be marked on the basis of the width and depth of the comments, discussion and analysis. Multiple-choice questions are usually marked right or wrong, a mark being awarded for a correct answer. It is very, very rare for marks to be deducted for a wrong answer, or for any marks to be awarded for a 'second best' answer, and therefore it is sensible to answer all multiple-choice questions even if the answer is a guess.

Practice

Examiners' reports point out that an examination in statistics always involves candidates in calculating some of the statistical measures which are specifically mentioned in the syllabus. **It is very important to practise these calculations so that the methods are fully understood.**

It is important also to have an idea of what the answer should be to any calculation. This means that by looking at a set of figures it should be possible, with practice, to have a good idea of the solution, that is, whether it will be around say 1, 10, 100 or 1000. This will make it possible to notice immediately if a large error is made in a calculation which leads to an 'impossible' answer.

Statistics examination questions

In preparing for an examination, an obvious query is 'what are the most frequently asked questions?' **Trying to predict or 'spot' questions in an examination is dangerous; it is fairly safe, however, to predict topics on the basis that these are the central areas of the syllabus.**

It is possible to take a chapter by chapter approach on examination questions to indicate in general terms the topics which are popular with examiners based on the experience of past papers. The areas that are not directly important for questions provide relevant background information. The table on page 235 summarises the position.

The General Certificate of Secondary Education (GCSE)

The aims of the GCSE syllabus are to produce courses which will enable the achievement of the acquisition of statistical knowledge and skills; the development of an understanding of statistical methods and concepts and awareness of their power and limitations; development of an awareness of the breadth of application of statistics and of an ability to apply and interpret statistics in everyday situations and in other disciplines.

The examinations assess a candidate's ability to recognise, recall and use statistical language and facts, perform relevant computations to appropriate

Chapter	Type of question	Type of examination
1	primary/secondary sources of data	GNVQ, BTEC
2	approximation (calculations), error	GCSE, GNVQ, BTEC
3	survey methods, sampling	GCSE, GNVQ, BTEC, RSA, I of M, Professional exams
4/5	presentation through diagrams and graphs	GCSE, RSA, GNVQ, BTEC, Professional exams
6	averages, particularly mean and median	GCSE, RSA, GNVQ, BTEC, Professional exams
7	dispersion, particularly standard deviation and interquartile range	GCSE, RSA, GNVQ, BTEC, Professional exams
8	probability, Z values, significance tests and confidence limits	GCSE, GNVQ, BTEC
9	index numbers	GCSE, GNVQ, BTEC, Professional exams
10	product moment and rank correlation co-efficients	GCSE, RSA, GNVQ, BTEC, Professional exams
11	moving averages and trends	GCSE, RSA, GNVQ, BTEC, Professional exams
12	tests, scales, experiments	Degree and diploma exams

NOTE: The 'Professional examinations' include I of M, ICSA, ICMA, ICA, ACCA or ACA, IPMD.

degrees of accuracy, select, present and analyse data, reason logically and make statistical inferences. There are usually three papers, an objective test paper, short-answer paper and a structured question paper.

In the statistics examination

It is essential:

- **to answer the question that is asked and not the question that it is hoped would be asked;**
- **to read the instructions carefully to be sure how many questions have to be answered** and which (if any) are compulsory;
- to make a timetable for the examination so that the time available is used to the best advantage (e.g. if five questions carrying equal marks are to be answered over three hours, 35 minutes could be allowed for each question leaving 5 minutes for planning and checking);

- **to answer all the required questions;** it is particularly easy in questions requiring calculations to become stuck on the detail of the mathematics and waste time which would be better spent on the next question (i.e. it is easier to score 9 out of 20 marks on each of two answers, than 18 out of 20 on one);
- to make plans for questions where necessary; statistics questions are often well structured and planning can be minimal, but plans can be used for deciding which questions to answer and which to leave when there is a choice, and ensuring a good and logical coverage of points in questions that require comment and discussion;
- **to decide exactly what the examiner wants in the way of calculations, facts, figures, diagrams and comments.** It helps to look for key words such as: describe, outline, analyse, calculate and so on, and to follow the instructions so that the question actually asked is fully answered.

Index

accuracy 12–25
  spurious 24–25
addition 21–22, 219–223
algebra 227–228
analysis of variance 196
approximation 12–13, 43
area table 216
arithmetic mean 92–99
arithmetic progression 205–206
association 160
averages 92–110
  arithmetic mean 92–99
  geometric mean 109–110
  harmonic mean 110
  median 92–106
  mode 92–96

bar charts 68–71
  component 70
  compound 70
  horizontal 69
  multiple 70
  percentage 70
  simple 68–69
bias 19, 33
  interviewer 33
binary 219
binomial distribution 192–193
birth rate 156–158
BODMAS 220–221
box and whisker diagram 105–106
break-even charts 84–85

calculators 8
cartograms 76
Census of Population 7, 35
Central Limit Theorem 38
checklists 198–199

chi-square test 195–196
class interval 55–57
classification 55–57
coding 28
coefficient of correlation 164–168
coefficient of variation 127
compound interest 206–207
computers 8–9
confidence limits 144–147
continuous variables 55–56
correlation 160–170
  Pearson's 165–167
  product moment 165–167
  rank 167–168
  spurious 168–169
  tables 164–165

data 1–8, 52–53
  primary 5–7, 26
  secondary 5–8, 52–53
death rate 156–158
decimals 218–219, 223–225
decisions 130–147, 209
degrees of tolerance 13–14
dependent variable 77–79
deviation 112–128
difference of means test 194
discounted cash flow 207–208
discounted rate of return 208–209
discrete variable 55–56
dispersion 112–128
  coefficient of variation 127
  interquartile range 116–118
  mean deviation 118–128
  range 114–116
  standard deviation 118–128
  variance 127
distribution of sampling means 140–141

distributions   113–128
  bell-shaped   113
  bi-model   113–114
  J-shaped   113–114
  normal   113, 119, 136–140
  rectangular   113–114
  skewed   113–114
division   21–22, 219–223

electronic calculators   8
error   14–24
  absolute   18
  biased   19
  compensating   19
  cumulative   19
  relative   18
  standard   37
  systematic   19
  unbiased   19
estimation   130
examinations   232–236
experiments   189–192
extrapolation   172

Family Expenditure Survey   7, 48, 155
financial (mathematics) statistics   205–209
Fisher's exact test   196
forecasting   171
fractions   221–223
frequency curves   65–66
frequency distributions   57–59
frequency polygons   63–64

Gantt charts   83–84
geometric mean   109–110
geometric progression   206–207
graphs   77–91
  semi-log   80–82
  straight-line   82–83
Guttman scale   203–204

harmonic mean   110
histograms   57–64
hypothesis tests   142–146

imperial system   224–225
independent variable   77–79
index numbers   148–155
  chain-based   153
  Laspeyres   152
  Paasche   152–153
  weighted   150–153
Index of Retail Prices   155
inductive statistics   192–196
inference   37
information technology   209–211

intercept   181–184
interest   205–207
  compound   206–207
  simple   205–206
interpolation   117
interquartile range   116–118
interval scale   229
interviewing   31–33
  formal   32
  informal   32
  interviewer bias   33
  respondent   32–33

Kolmogorov–Smirnov Test   195

law of the inertia of large numbers   38
law of statistical regularity   38
least squares   179–182
Likert scales   202–203
line charts   60
linear trends   179–185
  least-squares   179–182
  semi-averages   180
  three-point   180–182
logarithms   9
Lorenz curves   86–88

Mann–Whitney test   195
map charts   76
market research   196–198
mathematics   218–231
mean   92–99
mean deviation   121
measurement, levels of   228–230
median   92–106
metric system   224
modal class   94–96
mode   92–96
models   210–211
multiplication   21–22, 219–223

net present value   207–209
nominal scale   228–229
non-parametric tests   194–195
normal curve   113, 119, 136–140

observation   29–31
  direct   29–30
  mechanical   30–31
  participant   29–30
  systematic   30
ogive   92–96
ordinal scale   229

panels   50
percentages   225

perception   89–90
pictograms   74–76
  comparative   75
pie charts   71–74
  comparative   73–74
poisson distribution   192–193
population   20
powers   226
prediction   161
present value   207–209
presentation   52, 89
prices   3–4
primary data   5–7, 26
probability   131–140
product moment coefficient of
    correlation   165–167
proportions   227

quality control   145–146
quartiles   103–106
questionnaires   33–35
  design   34
  postal   33–34

random numbers   40
random samples   40–43
range   114–118
rank correlation coefficient   167–168
ratings   198–199
rating scales   201–205
ratios   226–227
regression   170
reports   28, 57
residual factors   179
roots   226
rounding   16–18
runs test   194

sampling   35
  bias   41
  cluster samples   46–48
  design   39–40
  error   38–39
  frame   40
  interpenetrating sampling   49
  master samples   49–50
  multi-phase   48–49
  multi-stage   47–48
  objective   37
  panels   50
  population   35
  quota   44–46
  random route   43
  replicated   49
  simple random   40–43
  size   39, 141

sampling continued
  stratified   43–44
  systematic   42–43
sampling distribution of the
    means   140–141
scaling   199–204
scatter diagrams   162–164
seasonal variations   174–178
secondary data   5–8, 52–53
semantic differential   204
sign test   193
significance tests   142–146
significant figures   17–18
simple interest   205–206
skew   113–114
slope   181–184
socio-economic classification   197
spreadsheets   209
square root   226
standard deviation   118–128, 140
standard error   140–142
standard normal distribution   137–139
statistical proof   186–189
statistics
  abuse of   2–4
  definition of   1
  descriptive   4, 37
  inductive   4, 37
  secondary   7–8, 28
  use of   4–5
strata charts   76–77
student's t test   193–194
subtraction   220
surveys   26–29
  design of   28
  objectives of   26–27
  pilot   28
  systematic   26–27
symbols of mathematics   230
symmetrical distribution   113
systematic observation   30

tables, statistical   28, 53–55
tests, statistical   193–196
  chi-square   195–196
  difference of means   194
  Fisher's exact   196
  Kolmogorov–Smirnov   195
  Mann–Whitney   195
  Student's t   193–194
  Wilcoxon   194–195
  Wold–Wolfowitz runs   194
three-point linear estimation   180–181
Thurstone scale   201–202
time series   172–184
  irregular fluctuations   179

time series *continued*
  moving averages   173–177
  residual fluctuations   179
  seasonal variations   174–178
tolerance   13–14
tree diagrams   135–136
trends   171
truncation   17
two-part linear estimation   180
type I and II error   143

variables   55–56, 77–79
  continuous   55–56
  dependent   77–79

variables *continued*
  discrete   55–56
  independent   77–79
variance   127
  coefficient of variation   128
vital statistics   155

Wilcoxon test   194–195
Wold–Wolfowitz test   194

'*x*' notation   187–189

Z chart   85
z values   138–139, 216

Poems

ℬ

The clemancie of Elephants. How elephants
breed and how they disagree with Dragons.

– A.B. Jackson

Owen Sheers
from Pink Mist

from *Part Five – Home to Roost*

ARTHUR
I don't remember any of what happened.
Just those howls, like dogs, as we drove out.
The fields and trees all black and green.
Some of the very first rounds, maybe.
But nothing else.

I had to pick it up all second hand,
as my hearing came back in the chopper,
and then again in Bastion.
How when my driver reversed
he'd hit a roadside IED.
How the explosion hit a fuel tank, or ammo tin
right under me.
Shot me out, like a jack in the box,
60 feet. And then how it all kicked off.
Rockets, grenades. The lot.

They took me straight to Rose Cottage.
A special room in the medical centre
deep among the tents and containers of Bastion.
A room for the lads or lasses who'd taken a hit
even the surgeons on camp couldn't fix.

It was manned, back then, by two blokes,
staff sergeants Andy and Tom. It was them
who took me in, off the ambulance,
and into their room. It smelt of sweet tea.
'That scent,' Andy said to me. 'It's the Eau de Toilette. Rose.
The Afghans insist we spray it on their guys.'
'Don't worry though Arthur,' Tom added on my other side.
'You'll soon get used to it. We did.'

And then they laughed. Not for themselves
but for me, I could tell. And they carried on talking too,
chatting me through all they'd do,
as they put what they'd found of me onto a shelf,
saying 'sorry it's so cold Arthur',
which it was, like a fridge.
Then they said 'sleep well' before sliding it shut.
My first night of three in Rose Cottage.

I saw them again just before I left.
When they slid me out into the light,
still passing the time of day
as they placed me in the coffin
that would carry me home.
Always calling me by name.
'Not long now Arthur.'
'You'll be back in no time.'
Gently, they lowered the lid
then, like two maids making a bed,
they unfolded, smoothed and checked for snags,
before draping me in the colours of the flag.

Jon Stone
Terrifying Angels

Tengu are roosting in our gardens,
in tilled shadows. Great slumped blackbirds.

Some have only one eye, or a tholepin
where an arm might foppishly dangle,
while some drag on slender pipes, exhaling
a smoke that crawls into the brain.

We cannot tell the males from females.
Our bed posts are their plucking posts
and in their sleep, they dream
they're up there, panicking the seraphim.

When we shamble to our kitchens
and draw together the tools to make coffee
(with a patience bordering on nobility),
still more of them gather at the glass,
their noses fat as plum tomatoes
and starry with pores.

But it's the shadows we shrink from:
our own shadows, always longer than us
and slung about our feet
like swelled and stepped-from skirts;

the shadows in our gardens, turning
slowly as an ache, as if they were gears
in some drawbridge mechanism;

and the tengus' shadows, of course,
that gather and close around us
like muscles in a gullet.

Allison Funk
Wonder-rooms

Though ordinary in their own habitats, introduced to one another,
  alligator to polar bear, ostrich and starfish,

they became a bestiary like none other on earth. A country
  found nowhere on maps.

The tusk of a narwhal, a dodo bird, mermaid's hand. More dream
  than museum, each wonder-room

hermetic as an ark where a collector could travel to his own Interior.
  Room within rooms, within.

Kingdom, Phylum, Class, Family, Genus, Species.
  And so I gathered them along a roadway in France:

Animalia, Mollusca, Gastropoda, Hygromiidae, Cernuella, Virgata
  clinging to a blade of grass

or in clusters sometimes, small versions of the grapes ready for picking,
  though not the Syrah's blue-rouge.

Pearly-grey instead, these slightly flattened globes. Their orbits
  inscribed upon them.

They're only *limaçons*, said a friend. Common vineyard snails.
  But my room has become the field

they came from: a Milky Way studded with them.

A.B. Jackson
from Natural History

Of Elephants

The clemancie of Elephants. How elephants
breed and how they disagree with Dragons.

How they make sport in a kind of Morrish dance.
How in drinking they may swallow down a horsleech

(which worme they begin now to call a blood-sucker).
The elephant who wrote Greeke and read musicke.

The elephant who cast a fancie and was enamoured upon
a wench in Egypt who sold nosegaies and wickerishe.

Their hornes, or properly Teeth, of which men make
images of the gods, fine combes, wanton toies.

Who march alwaies in troupes. Who snuffe and puffe.
Who the troublesome flie haunts.

Who cannot abide a rat or a mouse. Who are purified
by dashing and sprinkling themselves with water.

Who, enfeebled by sicknesse, lie upon their backes,
casting and flinging herbs up toward heaven.

Who adore and salute in their rude manner that planet,
the moone.

Paul Batchelor
Pit Ponies

> *But what is that clinking in the darkness?*
> — Louis MacNeice

Listen. They're singing in your other life:
'Faith of our Fathers' sounding clear as day
from the pit-head where half the village has turned out
to hear the latest news from underground,
news that will be brought to them by the caged ponies
hauled up and loosed in Raff Smith's field.

Could that have been you in Raff Smith's field,
mouthing the words and listening to the cages lift,
watching sunlight break on dirty ponies,
noting the way, unshoed for their big day,
each one flinches on the treacherous ground,
pauses and sniffs, then rears and blunders about?

As at a starting pistol they gallop out
and a roll of thunder takes hold of the field –
thunder, or else an endless round
of cannon fire. Hooves plunge and lift.
They pitch themselves headlong into the day:
runty, fabulously stubborn pit ponies.

But they seem to have a sense, the ponies,
a sidelong kind of sense about
getting called in at last light on the fifth day:
they seem to know the strike has failed,
as though they felt the tug of their old lives.
They shake their heads. They shy and paw the ground.

Let's leave them there for now, holding their ground
for all its worth, and say the ponies
might maintain their stand-off, as the livid
shades of miners might yet stagger out
of history into the pitched field
dotted with cannonballs you see today.

Let's pretend you might come back some day
to wait a lifetime on this scrap of ground
until the silence – like the silence of a field
after a battle – breaks, and you hear ponies
buck and whinny as the chains payout
and once again the rusty cages lift...

As though the day was won, as though the ground
was given, the ponies gallop out
to claim their piece of field, their only life.

Christian Ward
The Pony

Eccentric animal lover moves pony into semi-detached house

The pony is sitting in her favourite chair
looking at the Daily Telegraph. Human words
ricochet in her brain. Now she is crunching
on boiled sweets, remembering the field

developing like film in her mind. The pony
studies the room and sees grass. The furniture,
carpets, wonky lamp and TV are lush
grass stalks. Even the woman coming in

with a tray of tea and apples. She wants
to chew every inch of her while remembering
the endless field that kept her behind an electric
fence, away from thoughts of the road that rushed
quickly. Away from the thought of a thought.

Angela Topping
Sparrow

Passer, deliciae meae puellae – Catullus

When wind and earth joined together
to make the sparrow, they set
its toy heart flickering,
its small feet clicking. The breast
was made from speckled foam,
the wings painted with colours
left over from other creations:
burnt sienna, cafe latte, sludge.

Although the bird's beauty
was doubtful, it could weave in
and out of hedges, eaves and thatch.
The voice was nothing special:
a chirrup like a giggle fastened
in its throat like a comedy brooch.

Wind and earth baptised their child.
The first fairy godmother named it *passer*,
the second gave it joy, the third
the greatest gift of all: to be convivial.
The sparrow was a great success,
beloved of a poet's paramour, able to
hop into human habitations unafraid.

Wendy Videlock
Little By Little

Little by little the sparrow flies
toward the groaning epicentre
of a grand and historic twister,
and tumbling out on the other side,

silent as the artist's pride,
discovers the witch has already been squished,
the tornado story before has been told,
and the three wishes already wished.

Fate-ist! cried the sister finch,
Church-ist! keened the darkling thrush,
Mythist! cackled the squished witch,
the changeling, and the angel of reeds,

who'd broken their bones and left their souls
in the mulberry limb, and the watering hole.

Andrew Pidoux
Working Holiday

My head still uncracked by dawn,
My belly hollow and eyes encrusted,
I would creep out of the farmhouse
Into the chicken-patrolled yard
And make for the milking shed,
Where the farmer would be immersed
In the muscular dream of his job.
"That a boy," the farmer would say.

Then the cows crashed in wearing stilettos,
Drunk on their own swollen reserves.
From the pit that ran alongside,
I slapped on sucker candelabras,
And, one by one, drained the undersides,
The rich white shadows of these
Endlessly productive, childless beasts,
While inside, my family dreamed of sheep.

Matthew Sweeney
The One-eyed Philosopher Of Katmandu

The more a man wants, the less he gets –
so said the One-eyed Philosopher of Katmandu
over hot goats' urine and baked goat turds

that spring day I bumped my mountain-bike
up those steep, twisty roads to his green hut
camouflaged with lurid, sticky green creepers.

A bespectacled parrot was his sole companion,
and it chanted at regular intervals: *Death comes
to those fools who are never expecting him.*

The philosopher took out his glass eye to rub,
then replaced it in its socket. Typical of a bird,
he said, to be so sure of great death's gender.

I myself, he continued, know nothing about it,
beyond the fact it's a clifftop, and we all must
take the bewildering step off the edge into space.

And you, my young poet, he said, addressing me –
Be sure in your scribing to speak of that space
and nothing else, and then you may get everything.

Helen Ivory
Bluebeard The Chef

You coax the rabbit from its skin,
cradle the bruised flesh ripped with shot.
A deft incision and soon the tiny heart
is in your hand, its stillness
opens up a dark hole in the sky for you.

You climb inside
and all the stars are dying eyes
fixed into you like pins.
So you slice each optic nerve
and disappear.

The knife completes your hand
with such sweet eloquence
you part recall its amputation
when you were wordless
in your father's house.

Peter Bland
Echoes Of Empire

for my daughter

In Dad's old council house we sat
on sawn-off elephants' feet.
We slept in tropical hammocks
and chewed raw rhino meat.

Spears bristled in the hat stand,
Dead snakes were nailed to the stairs.
The sofa was always neatly stuffed
with old hippopotamus hairs.

My sister wore a wild grass skirt.
My brother bred tropical fishes.
Mother served us meals-on-leaves
and Grandma ate the dishes.

Father hid in the garden,
digging a pit for stray sheep.
He liked to use them as bait
for the tigers that walked in his sleep.

Lorraine Mariner
Cities

Once in Leicester Square
a bird wing touched my face.
It was just a London pigeon
whose spatial awareness
had gone out of kilter
but there was poetry in it.

And once on the Tokyo subway
on my way back to the airport
a Japanese commuter
wished me a good journey
home. How could I not
write such a thing down?

Shazea Quraishi
The Mummy of Hor

In this cave-like room, lamp-lit,
the Goddess Isis spreads her wings
across Hor's chest to protect him.

But that's not all:
the four sons of Horus guard his entrails
and the human-headed God Imset guards the liver
while Ha'py, with a baboon head, guards the lungs.
Duamutef, who has a jackal's head, guards the stomach,
Quebehseneuf, with a hawk's head, guards the intestines
and other Gods watch over his body
while sacred symbols protect his soul.

Hor's body, wrapped in layers of linen
and bound with black pitch
is here
and you are gone.
        I think of you
        on that country road
        when your heart stopped
        and your breath stopped...

        I think of you there alone.

Carolyn Jess-Cooke
Boom!

There was this baby who thought she was a hand grenade.
She appeared one day in the centre of our marriage
– or at least in the spot where all the elements of our union
    appeared to orbit
and kept threatening to explode, emitting endless alarm-sounds
      that were difficult to decode.
On the ridge of threat, we had two options.
One was attempt to make it to the bottom
of the crevasse slowly, purposively, holding hands. The other
    was see how long we could stand there philosophising
    that when she finally went off we'd be able to take it.
But then the baby who believed she was a hand grenade
    was joined in number: several more such devices entered our lives.
We held on, expecting each day to be our last. We did not let go.
As you might expect, she blew us to smithereens.
We survived, but in a different state than before: you became
    organised, I discovered patience, shrapnel soldered the parts of us
    that hadn't quite fit together before. Sometimes when I speak
it's your words that come out of my mouth.

Susan Wicks

Warning: Screen Will Not Stop Child From Falling Out Of Window

if you have a child, and the child's strong-willed enough
to drag or push a heavy chair across the slate
and pull itself up, and has the strength of wrist
to trip the clasp and turn the handle, the contrariness
to batter the mesh to bulges with its fist
or tear it with its fingers till it makes a jagged space

or if you yourself should get up from your desk
to let in the air and sunlight, birdsong, the scribbled shapes
of leaves, and the child should start to throw itself about
and scream, drive you half-mad with its impossible demands,
refusing to eat or sleep till you start to cry yourself
and finally bend to scoop it up and toss it out

or if the child were deep in some exacting game
with space-hopper-scooter-pogo-stick, and then took off
straight through the window with a look of mild surprise
to stand and brush the grass and needles from its knees
and walk away, not caring, from the chair, the pain,
the child-shaped hole that will let mosquitoes in.

Anja König
Advice For An Only Child

Go out to play they say – but you
can sit in the tidy garbage shed
until it's time to go to bed.

In the suburban summer the dozing roads
unglue – scratch off the plastic tags
and smell their toxic backs.

When you see a matchbox car
left in a sandbox, take it.
Run it against your bedroom door.

When they are loud downstairs go
to the bathroom with a book. No one
knows: a coin will open any lock.

When someone angry and grown up
says *you are not even human,*
say: *I am a little robot then.*

At night look at the albums in the attic.
You will find in black and white: a bald girl
in her smock, four boys in sailor jackets.

Adam Elgar
from A Day At The Zoo

Spider Monkeys

Twenty-fingered.
Masked like the muse of anxiety.

She wonders if they tried out being human once
and spend their days recoiling from the horror.

He takes the hint, thinks off her clothes, and coats her
in fine silvered fur with a flamboyant coil of tail.

She on the other hand can see the simian
beneath the skin without undressing him.

Gill Gregory
First Steps

Like a flower in a dream
toppling turquoise
and gold in the darkness

or a child as she falls
upon words
and their plight.

Tara Skurtu
Memory

You left the way women give birth.
The earth beneath swelled,
cracked and lured you toward
its sweet-lipped smile.

Like looking at an old scar
scab itself up then run red again.
Blood drawing back into a breathing
wound, skin yawning closed.

Sometimes, when I'm asleep,
you crawl into my ear, swim through
my thoughts like words being written,
the ink still shining.

Anna Woodford
Second Sight

Lying in the unforgiving morning on a low bed
(a year later – packing up –
I found I had been lying on nothing
beyond the rented mattress
but books and old newspapers)
as he lets himself out the inner door
of my flat then down the flight of stairs
with their threadbare carpet. Listening
in my pit for the final
click which has been coming
since the beginning of our spiralling
arguing. Still my mobile will call
me to call him – to take his leaving back.

A week or so later, I do not know
my life is holding
you in a corner of the bar
where I am being out the house.
Had I seen you cradling our baby
among the pints and dated clouds
of smoke, I might have done
something to upset the delicate
balance of our meeting which
as it happened, happened
right although all the way home I missed
the night's enormity – crying (laughably
now) on my brother's lurching shoulder.

Joe Dresner

C–

We could do with knowing what things like naves and balustrades
are and how we might go about fabricating them. Then we could learn
the properties of common minerals and their names, antimony,
bauxite, tantalum, which fill the mouth like the names of ancient
kings. We have spent too long coaxing out the inverted archaeology
of the heart or something like that. We stammer and our throats
close, our hearts flutter like semaphore. Pharaoh still lives. He sleeps
in his tomb, and when he dreams he dreams of his legions of fleet
footed cavalry and fields and fields and fields of wheat.

Hayley Buckland
Education

As Bessie leapt from the land
-of-take-what-you-want

into the long shiny oaked corridors
of goodness at St Claire's,

she paused a while to catch a breath.
Taking what you want is all

very well but the bags are heavy,
the handles, or is it the arms,

are taut. The sunset is of course
pretty but equally stretched

thin in parts, and on she bounds,
hamstrings barking at quads,

the reins tight on the arse,
and somewhere on the horizon,

some women are dabbling
at life, oblivious to the smoky

smells of s'mores and pubescent
girls asking about God and stuff.

Carole Bromley
Ghazal

I love talking to stones, love it, love it,
they may not listen but I rise above it.

Talking to a river's too easy somehow,
less of a competition. Stones give it

their all; rivers, on the whole, just chatter
(but don't tell them that, they won't have it).

If there's a point you want to get across
to a stone, don't bother, save it

for when you're talking to a river.
Never interrupt a stone; they'll not forgive it,

they'll say you talk too much then take
your share of the time and halve it.

Rivers won't fight, but show them an obstacle,
they'll move it

while a stone thinks its viewpoint's all that counts
and will stand and prove it.

But, Carole, talk to the stones all you like;
the river will hear you. Believe it.

Philip Gross
Variations On A Theme From The Cornish

An lavar coth yu lavar gwîr
Bedh darn nêver, dhan tavaz rê hîr
Mez dên neb dawaz a gallaz i dîr.

The old saying is a true saying
Never will good come from a tongue too long
But a man without a tongue shall lose his land.
– Cornish proverb

The man without a tongue will lose his land
Or: one who's lost his land will hear his tongue

grow stranger than the speech he moves among.
The land where loss is dumb will lose the man

and gain a stranger, and the stranger's song
will shift where few and fewer understand

lost to itself. The tongue without a man
will make a land of dumbness, that no one

who learns its lexicon can leave again,
its rivers swollen with unspoken wrong

and all the bridges down. Belong, be long:
the last words. They revert to wind.

A stone slip on the wet scree. Rain.
A lost thing at the door. A gift of tongues.

Stewart Sanderson
Fios

> Stram Yulloch. *A battle; a broil; given as syn. with* Stramash. Gall. Encycl.
> *This must be viewed as a variety of Strameulleugh.*
>
> – Jamieson's Dictionary

Fios means knowledge, which you don't have yet
I said to me in the smirr on Boxing Day
as the smirr came thick and blue, by the railway line:
inverted clouds off the fields and the nearly night.
No *fios* of knowledge yet, and almost dark.

Replying, a dog barked dangerous from the gorse
grown blue around the line. Did a shadow move?
I asked to myself when a pheasant clumsied away
through the asking smirr. Was it me or did something move
to take its *fios*. That something wasn't me
or wouldn't be if it was but I hadn't the *fios*
to caa through smirr and see. *Stramyulloch* was
an emblematic word for my lack of *fios*
as it traced a crazy cranreuch on my skein.
It could be that's the spelling: Scots, no further *fios*
required beyond that to get thick at the truth and the *fios*.
Or it could be that it's not. It could be *stramulleugh*
and a Gaelic loan – real evidence of *fios*
if accurate. It could be *stramulleuch*
or *strameuleuch* – or fifty other words
and that before the meaning's even touched.

So *fios* stands for knowledge, which you haven't got,
    I said to me in the smirr on Boxing Day
        while a dog howled hunt. No *fios*.
           You're in trouble now.

Martha Kapos
The Private Life Of The Tongue

At home unfathomable and secret
it had reasons of its own
tucked into a single bed behind closed doors

in the individual loneliness of his mouth.
For long periods in the afternoons
it would contract itself and draw in

unable to lengthen, reach over and open the curtains.
Did the room have a frescoed ceiling?
Attendants bowing and sliding backwards

in anticipation of *le grand lever*?
The word she wanted to hear
speaking to her alone: she imagined it

on the shiny tip of his tongue,
a mahogany piano standing upright in a corner,
superior and dark, its great lid shut

as Mozart waited lightly on the keys.
In all probability it had no more
to do with her than a chapel in the woods

whose altar with its one dry candle
had remained unvisited for years.
Never mind if it lay there dreaming

of the ritual of tucking in, the kiss goodnight.
She knew it would settle back to the shapes
it had taken in its other lives remembering

how it played havoc in a young boy
lying with its steaming pink root exposed
on a large plate in the kitchen

or hanging out a very red red
from the mouth of a black dog
it was impossible to avoid on his way to school.

James Aitchison
A Gap

A missing word feels like a missing state
of mind. Not absentmindedness: a gap,
a mental void I don't know how to fill,
a here-be-monsters on my neural map.

As a schoolboy I would turn a tap
and spates would spill
from mind to pen to pages of foolscap.

Now? Days and night and years. I can't distil
a sentence from the gap; no art or skill
can open the gates of light. I've made a date
with darkness once again. And so I wait.

Philip Knox
Northleach Brasses

> At least seven examples are known
>   – Douglas Gray, 'Middle English Epitaph', *Notes and Queries*

I had read in some article or essay
that Northleach Church, the precise
black rectangle of its profile like an unlit doorway
almost visible from your weekend place,
held an old brass carving with a seven-line slice
of a poem I'd once studied in a book of lyrics,
just the last, most hopeful stanza (and the least poetic,

a late addition tongued on in the plague years
when there was need, I suppose, for optimism).
I remember the pamphlet we found, a rare creature
hibernating amongst the church bulletins,
Northleach Brasses, the dreamt beasts its front was covered in,
a barking pig, an eagle strutting like a gull, a snail.
We already knew, weaving up the trail,

that our real purpose was the walk, the afternoon,
being together and out of the town,
before even seeing the church's bedroom gloom,
how half the place was tucked snug under shrouds
for restoration. It's in the looking-for that things become unfound.
And we could always check it in *Northleach Brasses*,
the photocopied rubbings all blotched in patches,

a prone couple boxed by that half-forgotten text.
Board-straight, hands clasped, they looked like lovers
lying pissed-off in bed (I fall asleep without regrets
like that with you sometimes, in revenge, cover
my face with sleep's plaster, withdraw the offer
of my five senses like the antlers of a hurt snail).
And transcribed opposite, the epitaph's consoling spiel:

Farewell my frendes, the tyde abydeth no man,
I am departed hense and so shall ye,
then all that shite about blood and redemption,
more cheerful than dust but no good to me.
For us, sleep's the closest thing – without dawn's breathed
soft forgiveness and forgiving soft flesh,
selfish sleep that lasts forever, stubborn as brass.

Kate Potts
Thirty-three

Now all the boys I've loved are married off, ensconced.
They bide in milky, clean-hewn terraces, in replicated seaside towns.

They wear matched socks. They wash. They see their own fathers' chins
and petulance – the kindnesses and tics – grow strong

and coarse in them, and this is comfort. They lullaby
their round-faced wives in lusty, baritone, newsreader voices.

Pour me a slug of this late August clarity of light: the contrast turned up high –
blunt as bone, acerbic as our windfall apples.

The garden's overrun with teetering foxgloves, cigarette ends, soup tins,
broken televisions; luscious, hoary, interloping weeds.

A fat fox grazes the rubbish sacks. Cars lope, tacit, by the kerb.
I hold my breath in tightly and bless the motoring

wish, wish of my pulse. On TV, the newsreader speaks of riots. His voice
is muffled pips and swells – is someone underwater.

John Wheway
Out Of Joint

After our lukewarm summer,
an unseasonable heat –

insects that were fading fast come
back to life. This bluebottle booms

among our bedroom's shadows;
staggering down the curtains, five wasps

get caught in your hair. Look how we leap,
night-wear tangling our limbs, our arms

thrashing round our heads. In this dim light
you might think we're trying hard

to dance. Only our swipes
with sections of yesterday's news,

our improvised swoops with upturned
cocoa cups, give us away.

Siriol Troup
Mashrabiyya iii

His soft looks,
his yellow slippers –
Come, he says, and she
comes.
But who has taught her this? –
To lie like a dove in his bed,
to draw her own reticulum
between them,
its jaggery of diamonds
grating at her heart.

Mashrabiyya: woodwork grille covering windows
of women's quarters in oriental houses

Vivek Narayanan
In-law Of S.

My brother: brother though brother made by human design and no
especially deserving biology; choice indeed though

certainly not my own; jig-dancing alien buffoon
manifesting one day in our living room,

so bizarre and yet – how could it be? – so right –
accidentally you completed us better than fate

or custom, simple will or jiggered scripture (preservative
measures which somehow in the very stale and rebarbative

crust they harden turn equally unaccountable loss) –
My brother: the other hand: my pig- and cow- eating Walloon, my piss-

taking, law-book thumping, stock price memorising, labour
un-alienating, seat-reserving, liquor-loving wunder-

-bah: your alien speech, your hidden power reserves, your own
unuttered but unwavering rules, applied diligent in the midst of prone

and relentless chicanery – and our jagged nervous mugs, our shudder
upon the non-returnable admittance of a variety of other

species into this system so carefully conserved by habit and by caution and
perhaps almost pathological laws of modesty; then, our new man-

made rewiring of pre-rigged laws, a fresh and altogether
different idea of sacred: a breath by which to power

the zeppelin into treacherous futuristic skies thanks
in no small part to you. My brother, the thought sinks

in, Did we deserve you, you us? Evolution (aka the universe)
happily thankfully refrains from commentary on its many bursts

of improbable planning. The human in jigs and apparently foolish jags
reassembles; alien gets necessary, banal: affection, like an old dog's tail, wags on.

Mr S. is a playful alter ego in Vivek Narayanan's *Life and Times of Mr S.*, forthcoming from Harper Collins India.

Elizabeth Barrett
Persimmons

In the mornings my teenager's room
smells of them; the antiseptic tang
of semen hanging bittersweet on the air.

Autistic, he lacks secrecy or shame –
lies dazed on the bed, sheets pulled back,
or stumbles from his room, penis erect.

When he is away in respite care
I open windows, strip his bed. Tonight –
in the very dead of night – I slip inside.

There is a man in my bed I barely know –
we came here from an empty car park,
drove in convoy down misty roads.

I think of him alone under my sheets –
imagine him in the recovery position;
I turn his body to marble then flip him over –

make him dream and twitch. I stretch
my legs to the end of my son's bed –
catch the sour scent of persimmon

on my skin. I close my eyes –
try to guess the way he lies here in this bed –
whether pictures of me ever fill his head.

These first fruits: a sudden bloom of hard
flesh as it swells in our hands, blushing
and oozing, the sweet juice spilling.

Dan Wyke
The Platform

Here comes the pocket-timetable man,
beard matted, walking with the awareness of a meditator.

He holds the page up to his face, his glasses are so thick,
and walks to the end of the platform

where it tapers to a point, and back
just as far as the waiting travellers.

To and fro he goes, eyes down, approaching
and re-approaching the end like a diver on a high-board.

Richard Lambert
The Maid

With its deep, swirling carpets and uplit walls,
the wooden staircase, the *whsk* of metal
of the lift's concertina doors, its clanks and whirs,
slowly the hotel stirs.

The guests appear in ones and twos,
scented by aftershave, seaweed shampoos,
but drifting slowly, as ghosts might.

I strip the beds
possessed of each room's staleness, dust,
UHT milk tubs, empty coffee cups,
nostril hairs on pillowslips,
wrinkled tissues in a bin,
blood stains on bed linen.

In one guest's tidy, rumpled room
I touch a woman's cardigan,
lift her lipstick, carmine red,
beside a note to self – *scarf for X.*
I leave the room nearly as it was.

In one, I open windows.
The muslin curtain blows.

At night upon my narrow bed
beneath the rafters, on the wall
I see a picture of a lake.
Under moonlight, clouds, or rain,
across the city's
mansard roofs and gable ends
I sift the grainy light,
and weigh what's taken and what's left
to dawn, its sudden, total theft.

Michael Hulse
In The Peloponnese

At the next table, grizzle-jowled, silver-haired, in his undershirt,
before a proprietorial spread
of zucchini, tomatoes, peppers, an Amstel beer, and a napkin of bread,
taking his elbowed ease for a lonely and difficult day at sea ahead
now that the weekend Athenians have departed,
the owner of the Panorama restaurant's at meat.

Aleppo pines and cypresses and figs
gentle and tremble and make a shade
where a cat the colour of Pericles
on the old one-hundred drachma note
rhetorically purrs *what price the Euro?*
to an imaginary parliament of fowls.

The nets are drying on the wall,
the water talks of love, I think, but not of currency,
and nothing, believe me, nothing at all
depends or ever could depend, on earth, in heaven or in hell,
on the cherry-tomato Fiat Seicento
parked on the seafront before the Apollo Hotel.

Rosie Breese
Ascension

You were nearly complete
although some flesh still clung,
the eyes were fearful and the throat could cry.
The brain was closer. Its fault-lines
had opened to the sky
that flooded empty neural pathways,
bleached bone, melted muscle,
backlit the mask
hiding your transfiguration.

We tried to hold you down by your hands
but you slipped out of them,
shouting for us from behind your face
as you evaporated into God's cold heaven,
wordless by His perfect design.

Jane Yeh
The Ghosts

We summer wherever you are in invisible log cabins.
We gather under umbrellas whenever the sun comes out in force.
We hover above the ground because our feet don't work as normal.
Our low moans give us away when we try to sneak up on you.

Contrary to popular belief, we aren't out to avenge our deaths.
We don't need help locating scapegoats or generating confessions.
We follow current events by watching TVs in shop windows and pubs.
Sadly, our electric auras tend to interfere with the signals.

We cluster round fountains because they attract wishful thinkers.
We try to enjoy life even though we're noncorporeal.
We frequent local hot spots, boîtes, crêperies, and arboretums.
We also ride on the carousel plates that rotate inside microwave ovens.

If you look, you can catch us shadowing your children to school.
Our rucksacks contain miniature rucksacks for storing small objects securely.
Some of us own dachsunds because we don't have any siblings.
Some of us think death is just another kind of being lonely.

We linger near burger bars to regurgitate our youth.
Our pseudo-fluorescent presence blends in at chemists' and delis.
If you see us circling round drains, it's because discarded things end up there.
We don't mean to spook you, we just want to be noticed.

Julia Copus
Miss Jenkins

More and more, lately, when absence thickened the air
at the schoolgates, in the street, first thing on waking,
she'd think of her former calling, the way it had defined her.
In the dim, sugar-paper blur of the light,
while boiling the kettle or kneeling over weeds,
many times at dusk now (the streetlights coming on)
she'd feel herself alive, transported
once again to the bright, tall-windowed classroom,
chalky-fingered, cherished by her peers, and walking –
that brisk and rhythmic pace she adopted, all her working days.
Even in sleep, her breath would rise and fall with
the sharp pat pat of the children's feet approaching and
she'd sense – in her blood – like a counterpoint beneath it,
the slap of books upon each child-size table
whenever she set up class for their arrival.

Whenever she set up class for their arrival
– the slap of books upon each child-size table –
she'd sense in her blood, like a counterpoint beneath it,
the sharp pat pat of the children's feet approaching and
even in sleep her breath would rise and fall with
that brisk and rhythmic pace she adopted, all her working days.
Chalky-fingered, cherished by her peers, and walking
once again to the bright, tall-windowed classroom,
she'd feel herself alive; transported.
Many times at dusk now (the streetlights coming on),
while boiling the kettle or kneeling over weeds,
in the dim, sugar-paper blur of the light,
she'd think of her former calling, the way it had defined her,
at the schoolgates, in the street, first thing on waking –
more and more, lately, when absence thickened the air.

Angela France
Other Tongues

When I say I'm alone, I'm lying.
My mother tongue sleeps under my skin,
bred in the bone, colouring my blood.
I speak from an echo chamber
where the walls pulse with whispers,
familiar cadences rising and falling
at my back. I speak from a limestone floor,
as familiar to my feet as are the bones
of the hill creaking between the roots
of great beeches. I speak with multitudes
in my throat, their round vowels
vibrating in my stomach, their pitch
and tone stiffening my spine.

Maurice Riordan
The Navigator

for Michael Murphy, 1965-2009

Always the same seat, gliding backwards, north,
when the dawn broke my side of the Pennines
beyond Newark, where I looked out for herons
on the reddened lake and thought of you in Aigburth
patient in the sky-lit room, this same light
thundering at you like Handel's Philistines
even as the thought struck – how in seconds
it would flood your longed-for West and ignite
the ocean from Benwee to Erris Head...
Let me repeat a local's words, wind-caught
as the seas heaved and stretched to Reykjavik:
Wasn't he very brave, Brendan! She shouted,
Brendan the Navigator – to go out on that
without map or guide in his scrap of currach.

Gillian Allnutt
the shawl

In memoriam Julia Darling

of air and wool

her frail earthwhile

who promised her people

palanquin, purple and pall

and left them all

a little something fit for April

snowfall

Simon Royall
Putting On My Shirt

In the dark, I can still find each button
and so dress by the bed without waking you.
I find them like the stops of an accordion,
my fingers pressing silently, practising,
feigning musicianship to themselves,
since I never learned to play and moreover
in my own hands I am collapsed, silent,
folded all day, my full-bellowed length
kept compact – no tune getting through
my clumsy fingering. Only once I'm back
does instrument find musician,
do I commend to you each button.

Robert Hamberger
Shoelaces

Untying shoelaces late at night
I think this is how people live:
untying our laces, drinking with friends,
running on a wheel, fretting and laughing,
dipping our heads under a lilac branch
as it leans over a fence into the street,
hearing a blackbird on an aerial
welcome the evening, knowing we'll never know
when time will take us, but it will,
the lilac, the blackbird, the shoelaces
always unravelling, infinitely precious,
the moment we sit on the edge
of the bed, late at night, after a drink
with friends, thinking this is how people live.

Kate Miller
Again (*reprise*)

Rain set in at six, sea flecked
like wet tweed, heavy trousers, hoisted
with a webbing belt, sky a pale shirt which shed
its light. Bit by bit it fluttered into ash,
until almost everything was lost
from sight, when – and I did not appreciate
how far, the distance you had come –
you reappeared. Believing no-one
could be out that night in their right mind
I did not want to open then. And yet
I let you in. I would again.

ℬ

Tim Cockburn
To A Stranger In Company

Guessing what draws you, what draws me to you
is the shape your courage makes, now, from here,
and my faith, if you like, is that however distance
like grass should obscure it, it will betray,
as a scorched field in summer the site of a former church,
an impression at least of that patient construction.

Christopher Middleton
The Ghosting Of Paul Celan

In shadow from the past
I have tried to tell
of what the breath-crystal
in a word rejects:
greed of the eyes to see again,
greed of the fingers, all five.

Against the odds an image,
against the flow,
no pronouncements,
advancing –
                Thicket, the image
lunges through the thicket
out there, in the head.

Firebomb and
martyrdom, how
neighbourly in sound, by wires
people move, Punch
and Black-Eyes. Wires twist
round cortex and ankle.

Slow image, painful, breathless:
in ordinary civvies
four or five of them,
seen from behind, four or five
dispersed, walk forward beneath
the living branches.

In the red gaze of wine
in its house of glass
there'll be sometimes a dragon;
sometimes a reminder
that trust is for the free,
the foolish, the very rare free.

Sliding through cropped grass
boots and shoes

and then the foreknown
about to happen, any moment
some unspeakable thing.

See then ways they mix
into the mutations. If we cringe,
the god will spit on us;
apprehend the tact while pipe and string
carry us away,
and in the music we are lost.

Frank Dullaghan
The Crash

Everyone wants to be paid
but we have no money.
They call me. What can I say?
I see the wedge cut out of the trunk
as I stand in the tree's shadow.
It is tilting towards me.
I hear its pain, can feel
the snap and rip of its fibres
before they explode.

We have let the staff go.
They are angry.
We have not paid ourselves
in four months.
Everyone has battened down
in this financial storm.
We cannot close any deals,
our months of effort are like leaves
thrown into a wind.

I park the car, watch a cat
slink between a white wall
and a fence. The day
is bouncing in its cot
and wants to be brought
downstairs. I take the elevator up.
My desk phone is already screaming
across the empty office
when I open the door.

Terese Coe
More

The devil has more money than you think.
He's paying off the players as they drink.
From suicide, coyote, croupier,
from mayhem that's sufficient to the day,
he gets his cut.
From killer, con, and cheat –
the devil has more money than the Street.

Tom Warner
Wallets

When I consider the history of wallets
I picture men with moustaches and rigid hats,
the kind that knew the weight of money
and every week would count a modest sum
for pins and bits and bobs into a waiting palm.
Those men have tipped their hats
and climbed aboard their steaming trains
or simply scurried out of shot.

Some wallets have a window for a picture
of a wife or child (behind which one might slip
a number written in eye pencil),
and some wallets are laid like a gun
on the bedside table of a one-night-stand.

The first I owned was *Genuine Leather*
and smelled of the indoor market
where my mum bought the veg from loud men
with money bags who winked and called her duck.
That first one took some getting used to;
I walked out with one back pocket off balance,
learned to sit lopsided while it lost its corners
and grew supple as a foreskin.

Dan Burt
Homage For A Waterman

He jams his clam tongs down three feet
and fetches bottom, pulls them so wide
he leans spread-eagled, then scissors
back, heaving till his knuckles meet,
and hoists the bales over the side
hoping for little necks or oysters,
a black oil-skinned stick figure
pile driving in November sleet.

Townies charter him in season
to cast the bars with bright lead lures
or eels, a year's bookings complete
before it starts, and find him on
the dock at dawn, a white sweat shirt,
khakis, red and black plaid coat,
a Triton other skippers query
on where to run, what wrecks to skirt.

Thirty years ago time took his boat,
in due course him, then his parties;
where he rests, an urn, beneath a stone,
I never learned, while the boy
who begged to help his hero lift
the hook wears plastic knees for bone.
But when winds sough and sea gulls toy
with thermals by my hollowed cliff
his leathered face looms into sight
through spray and howl, helm held tight,
bow to seas breaking in the bight.

Geraldine Monk
Forgiving Mirrors
Abstractions Towards My 60th Birthday

1

When reflected the forms were devious
hovering on convex anarchy.
My mind slides through warm curlicues.
Brutal refractions. Time spaced-out.
Contracted.
Was it only yest..?

All our yesterdays wrapped in newspap
shocked a past tense world. Accumulative
weary. Non-identified humours run cute to
unfunny bone. Evaporated laughter seeps from
lost dances of earthly delight throwing shadow.
Flames.
Part and par..?

When excavated the skeletal forms were
furtively touching outlines of languid beauty.
Ghost bouquets pining for sun food and cave-
age prima donnas. To bury hooded thoughts.
Wild eclipses. Random blizzards. Shedding.
Loads.
Is that all..?

2

From Korea to Afghanistan – we can
measure our life span in wars.
Negotiating tempers slung reckless on
hairpin bends – nothing and all for
ever reversible. Pathetic breeze
struts the heart with momentary catch.

    circulationcirculation

Every seven years a different person
different versions of enduring same.

    circulationcirculation

This giggle game trickles to the brink.
This weeping reason breeds the mantis
dance amongst the ruins. Embers at
dawn. Hypnotic somnolence of spooned
molasses. You have to laugh. First post.
Last rites. Muddle through.

    circulationcirculation

Anthony Caleshu
If We Had A Map

From up here, we can almost see where we've been coming from.

We'd be off the map altogether, if we had a map.

If we had a map, we still wouldn't know how to get back.

Tap, we say with our boots on the ice. *Tap-tap, tap*.

To the explorers before us, we say we never meant to go exploring.

We never wanted to know the unknown.

Our lack of ambition was in our lack of attitude.

We dissed the converging of longitudes.

The longitudes have converged to this point where time is irrelevant.

But nothing is irrelevant.

Not us, not the sharks we imagine below us.

Latitudes circle us like sharks.

It would take some serious maths to compute the depths of our frostbite.

How many mega-bites of ice would it take to get home?

If we had a virtual bike, we'd put virtual chains around its vitual wheels.

With each step we take, we spin our wheels.

We're cycling into all the time zones in all the world.

We watch our kids waking up, our boss staking up, our wives taking their place
in this world without us.

Everything is South from here, even the light.

We poke a hole looking for some life on the other side of the world.

If we had a map, we'd twist it into a swan or a swan-boat.

With each step we take away from this Pole, the world gets smaller and smaller.

Our footprints are as small as our thumbprints.

With each step we take, the angle of tilt gets demanding.

What rotation of axis is needed to turn us home?

We have no passports and invite remanding.

Fiona Benson
Soundings

There's a leveret in the field.
I know it by its mother's haunt at dusk,
can sense the cupped space of its watch
over near the gorse.

For now it's stowed
belly to the thawed ground, screened
in timothy and vetch, tuned
to the wing-chirr of insects,

the far-off bark of a fox.
As for you, small one in my womb,
the midwife lies an ear down flat
to hear the wild, sweet beating

of your heart,
scans for tell-tale movements on the graph.
There are still so many ways
you could startle, abort.

Theresa Muñoz
Delete Where Appropriate

if only we   had a record
of everything we ever said
a silver typewriter
that clicked and published
our words and sighs   as they evaporated
if only we had a transcript
of our bad years   our better ones
a list of criticisms spat
across the kitchen
a register of kind phrases
uttered before bed   or in the car
my hand on your thigh
in the melting dark   if only we

had a database   a spreadsheet
we could search the terms
of our agreement
had we bothered   to write one
had we been honest
about ourselves   our wishes & failings
perhaps I wouldn't be
on the bottom stair
eyes pressed into my knees
conjuring looping screeds   of ivory paper
that I could amend   delete where appropriate

CENTREFOLD

ℬ

An education into regions of
double-jointed nonsense and joy...

– Adam Piette

Poetry Parnassus

SIMON ARMITAGE

Poets are the tramps of literature. That's both a complaint about the way we have to peddle our handiwork and a grateful acknowledgement of the freedom and independence associated with the role. We are itinerants of sorts, often on the move between readings and residencies, between one idea and the next. It's a portable art; I might have visited a few dozen poets in their homes but I've bumped into many hundreds of them in green rooms, bookshops, lecture theatres, conference centres, foyers and reception areas. When I gave up being a full-time probation officer to become a full-time poet I thought I would be trading the mean streets of Greater Manchester for the relative peace and quiet of a book-lined back-bedroom, but the multi-storey car park behind Wakefield Westgate Station has become a more familiar haunt.

In 2010 I began a residency at London's Southbank Centre, and announced my intention to try and allocate poets a space they could call their own, especially those poets from outside the M25 who often find themselves in the capital for one poetic reason or another but, not being members of the Athenaeum or the Groucho Club, end up wandering around in the rain or the traffic. I thought the Southbank with its central location, its dedicated Poetry Library, its refurbished public spaces, re-affirmed open-door policy and its long-standing relationship with the homeless could provide the ideal venue. All we need is somewhere to sit, read, talk, drink, log-on, think, even write, and on that basis I conducted a rather non-scientific feasibility study, wrote it up and handed it in. I'm still waiting for a reply. Daydreaming in a more abstract way about putting a roof over the heads of wandering poets I also suggested that in Olympic year we should invite a poet from every participating country for a kind of unprecedented global poetry gathering, and they said, "OK, let's do it."

Mount Parnassus in Greece is said to be the sacred residence of the god Apollo, spiritual dwelling of the muses and the home of the "first poet" Orpheus. During the last week of June this year it is our ambition to recreate the foot-slopes of that mountain along the shore of the river Thames and among the many rooms, bars, workspaces and auditoria belonging to the Southbank complex. For over eighteen months we've been attempting to identify poets from the two hundred and four Olympic countries (the number seems to vary week by week as nations emerge, dissolve, abstain etc), beginning with a long nomination and consultation process. Ultimately, though, this is a curated and programmed event, and we have only invited poets whose work excites us and whose presence we hope will bring vitality and integrity to the project, some decidedly literary, others from storytelling, oral or performing traditions, some world famous, others barely known outside their own borders (and some hardly known within them either). It probably goes without saying that this has been a mind-bending administrative and bureaucratic task with eventually many dozens of people working on the Parnassus project, overseeing the anxieties, aspirations, dietary requirements and airline seating preferences of several dozens of poets on an almost daily basis and in over fifty different languages. In more anxious moments I do worry that we have created an utterly unmanageable nightmare scenario, an amalgamation of the Tower of Babel and the Eurovision Song Contest. But I also cling to the idea that Poetry Parnassus might be unique, not just in its size and ambition (the biggest ever?), but in its attitude and ideology. In its *daring*, in fact. Building on the Southbank's Poetry International Festival, and with a nod in the direction of the landmark and legendary 1965 Albert Hall event which went down in history as "the Poetry Olympics", Parnassus has developed into a week of readings, translation, seminars, conferencing, workshop and discussion. As well as headline-grabbing events such as Casagrande's *Rain of Poems* (a hundred thousand poems in the shape of bookmarks dropped from a helicopter), *The World Record* (a book consisting of handwritten poems from poets of every nation) and gala readings involving Nobel Laureates, the schedule also includes more intimate or specialised events, reflecting the range of poetic voices at work in the world today, and recognising the varying forms and approaches poetry might take. Parnassus will also include launch events for five new collections (Bloodaxe Books will publish an accompanying anthology and *Modern Poetry in Translation* is dedicating an issue to Parnassus), a day-long conference or "summit" debating issues such as poetry and money, and a special event honouring Ted Hughes's involvement with both Poetry International and MPT, including readings from his letters and recitals of his poems in Greek, Turkish and Italian. Poets such as Seamus Heaney, Jo Shapcott, Wole Soyinka, Kay Ryan, Yang Lian and Jack Mapanje will already be familiar to some readers, poets such as Kazakhstan's Akerke Mussabekova (our youngest poet at 24) and Luxembourg's Anise Koltz (sixty years older) less so. Inevitably, the world being the place it is, voices of protest and survival will also speak loudly during Parnassus: Albania's Luljeta

Lleshanaku grew up under house-arrest, Jang Jin Seong was the former court poet to North Korea's Kim Jong-il who fled to China and then South Korea, Cambodia's Kosal Khiev was born in a Thai refugee camp and in 2011 deported from the USA, and Nicaraguan poet Gioconda Belli was a Sandinista revolutionary in the struggle against the Somoza dictatorship.

Poetry, we're told, was part of the original Olympics, and quite possibly an actual event. But in contrast with the great political and financial behemoth into which the modern Olympics have morphed, Parnassus was conceived in a genuine spirit of participation and cooperation. It was also conceived as non-competitive in the sense that there will be no podium positions and no use of the word medal in its verb form, though the extent to which personal glory and "winning" inform the contemporary poet's psyche might have been another of our seminar subjects. Neither will there be national anthems or any expectation that poets will drape themselves in the flags of their countries or dress in their national costume, either literally or metaphorically. All we're asking is that those invited bring a sense of their own poetic language to London for a week in early summer and use Parnassus as a platform for delivery and a forum for interaction. In my experience, poets tend to have complicated and conflicted notions of nationality and nationhood, and if Poetry Parnassus turns out to be an opportunity for poets to distance themselves from borders and frontiers or even speak out against them, so be it.

Translation will be the keystone of this project, and translation, as we know, is the great impossibility of poetry. Yet poetry would be a moribund, hermetic and sterile activity without it. Parnassus itself is a similarly impossible proposition, though if we do fall short of our original intention it will not be for want of trying or through lack of ambition. On that note, if you're reading this and happen know of or even happen to be a truly excellent poet from Brunei Darussalam, Madagascar, Monaco, Papua New Guinea, Timor-Leste, Vanuatu or one of the sixteen other Olympic countries for which we have so far drawn a blank, please speak up.

Poetry Parnassus is at the Southbank Centre, Belvedere Road, London SE1 8XX, from 26 June to 1 July. www.southbankcentre.co.uk

Poetry In The Prose: Getting To Know The Prose Poem

CARRIE ETTER

When I arrived in England in 2001, I tended only to find prose poems in more experimental journals such as *Shearsman* and *Tenth Muse*. Now I regularly see them in a wide range of literary magazines, and the first contemporary British prose poetry anthology has been published, *This Line Is Not for Turning* (ed. Jane Monson, Cinnamon Press, 2011), to positive reviews. The palpable increase in interest from fellow poets and students alike has been an exciting pleasure for me, as someone who began writing and publishing poetry – including prose poetry – over twenty years ago.

One of the fruits of the proliferation of prose poetry should be a greater pluralism. Monson's anthology has some unexpected bedfellows, placing the work of Richard Berengarten and Jeff Hilson alongside that of Pascale Petit and George Szirtes. Similarly, a special feature on British prose poetry in *Sentence: A Journal of Prose Poetics* brought together John Burnside, Rod Mengham, Geraldine Monk, Peter Reading and Peter Redgrove, among others. In fact, prose poetry seems to nourish styles that do not easily fall into the usual, tired distinctions of experimental or mainstream, as with Luke Kennard's surrealist narratives and Ágnes Lehóczky's psychogeography, as in her compelling poem, 'Prelude', gazing on a cathedral ceiling:

> To get to the core of the place they have been traveling to for so long to people an empty city, a city with no topography, the sky without impasses, cobbled cul-de-sacs, crowded catacombs, horizontal reminiscences. They travel so they can be exactly where you are now. They travel to settle, you say. To illustrate the biosphere around us. To illuminate the darkness tonight. They arrive. To live among us. Slow rows of caravans, bright lanterns, departing on the ridges of the vault. On the edges of the universe. Unclear. The difference. Between departures and arrivals.

As more collections include prose poetry, however, we face an important problem. Critical discussion of the form lags behind its publication, and consequently prose poems, in books primarily consisting of lineated poetry, often go unmentioned. When new volumes composed wholly of prose poetry appear, such as Linda Black's *Root* (Shearsman, 2011) and Lehóczky's *Rememberer* (Egg Box, 2011), they are less likely to be reviewed, and those reviews that do appear are more likely to neglect discussion of the poet's particular techniques.

In a moment of rare reviewing honesty, Paul Batchelor, at the end of his review of Simon Armitage's *Seeing Stars* (Faber, 2010), comments, "Are they poems, or prose poems, or flash fiction? I'm not sure." While some poets and critics insist that we must resist defining prose poetry for it to retain its subversive, genre-blurring character, I find some basic distinctions crucial for its appreciation. While a lineated poem's development requires some sort of progression as it moves down the page, most reductively a movement from point A to point B, a prose poem develops without "going" anywhere – it simply wants to inhabit or circle A. If the prose poem takes narrative form, that narrative operates to represent or suggest a single idea or feeling; the story or plot is there at the service of an idea. Otherwise the piece is a form of narrative prose, such as a flash fiction or an anecdote, rather than a prose poem.

To clarify this distinction between a narrative prose poem and a piece of narrative prose, consider Antony Rudolf's piece, 'Perfect Happiness', from *This Line Is Not for Turning*. The work begins with the announcement that the speaker is ten years old and has just arrived at his grandparents' house; he goes on to relate his activities over the course of the day: wander about, look at comics, throw a tennis ball against a wall, etc. The poem's momentum derives from this succession of events. The point of the poem, however, is not the story so much as the way these simple events add up, either in retrospect and/or as they are experienced, to an overall sense of 'Perfect Happiness', to that single idea or feeling. That quality distinguishes the poem from anecdote or flash fiction.

More than one regular reviewer of poetry has told me that s/he will sooner decline a collection of prose poems than cover it; if faced with individual prose poems amid lineated ones, s/he might address its content, but would feel wary of discussing technique. Yet lineated and prose poems share much technical ground: use of metaphor, repetition of sound (alliteration, consonance, assonance, partial rhymes, etc), imagery, and voice, just to start. The difference comes down to the sentence (and it may not be a complete grammatical one at that) rather than the line as the primary structural unit.

This means that instead of looking at a poem's line and stanza lengths' contribution to structure, we consider sentence and paragraph lengths as well as sentence types. For example, the succession of short, subject-verb sentences in Carolyn Forché's brilliant prose poem, 'The Colonel', enhances the dramatic tension with its staccato effect on the rhythm. In the delightful 'Hedge Sparrows', Richard Price conveys the bird's incessant chatter through one long, long sentence – of one hundred and eighty words! While these are more pronounced examples, they give a sense of the relationship between sentence length and structure and the poem's meaning, just as we would consider with a lineated poem's use of the line.

The more we engage with poetry's possibilities, the more poetry as a whole benefits from the exploration, and that engagement means thoughtfully developing critical approaches to each form that emerges: giving each new expression of language the attention its eloquence has earned.

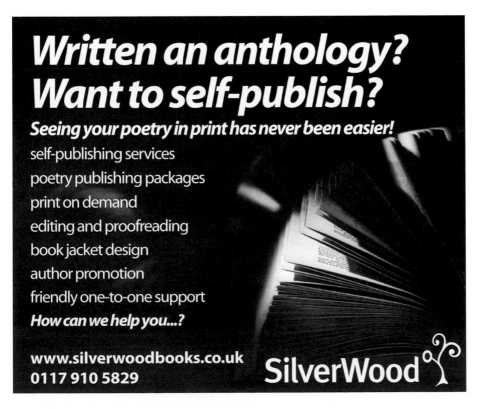

Beyond Dictionaries: Vahni Capildeo And Etymology

PETER ROBINSON

Geoffrey Hill, current Oxford Professor of Poetry, illustrates beliefs about poetic technique as a drama of sin and redemption in his essay 'Poetry and Value' by referring to "supporting evidence" that is "preserved in and by the *Oxford English Dictionary*". W. H. Auden, though no less a believer, suggested that "a dictionary is absolutely passive and may legitimately be read in an infinite number of ways", while Ralph Waldo Emerson thought dictionaries worth reading because they contain "the raw material of possible poems". Harry Guest's *Some Times* (Anvil Press Poetry, 2010) contains just such a poem, one called 'Divans', which riffs on a related bundle of meanings:

> Lounging on one
> in Persia or
> leafing through one
> to please or bore …
> Yes, *couch* or *set*
> *of poems* – yet
> it seems you call
> that *Customs-Hall*
> we scurry through
> in ones and twos
> with fags and booze
> a *divan* too.
>
> Poets can't claim
> immunity.
> The rule's the same –
> if duty-free
> that minimum
> demand's still fair:
> 'Have you got some-
> thing to declare?'

Recorded in English since 1586, meaning 'Oriental council of state', from Turkish *divan*, from Arabic *diwan*, a Middle-Persian loan-word in Arabic *dīvān*: 'bundle of written sheets, small book, collection of poems', the senses evolved through 'book of accounts' to 'office of accounts', 'custom house', 'council chamber', then to 'long, cushioned seat', such as found along the walls in Middle Eastern council chambers. Inspired by the *Divan-i Hafiz*, Goethe named his 1819 collection the *West-östicher Divan*, and Edwin Morgan jokingly called one *The New Divan* (Carcanet Press, 1977). Guest's poem finesses its close – perhaps prompted by Wilde's answer ("Only my genius") to the US customs official who asked their standard question – by suggesting that crossing borders, living in different countries and languages produces occasions for poetic speech, while the word 'divan' illustrates how there's frequently nothing Anglophone about the English language, it being a hybrid pidgin. Guest's poem is dedicated to the Jamaican poet and broadcaster A.L. Hendricks (1922-1992) whose experience of 'translation' from the Caribbean to London may add to Guest's implications, for, until recently, if words crossed borders and were absorbed in other language areas, it was because people transported them, as happened to the Persian ناوید, *dīwān*.

The much-travelled Trinidadian Vahni Capildeo's third collection, *Dark and Unaccustomed Words* (Egg Box Publishing, 2011), takes its title from a passage in George Puttenham's *The Arte of Poesie* (1598) opposed to uses of "all darke and vnaccustomed wordes, or rusticall and homely, and sentences that hold too much of the mery & light, or infamous & vnshamefast" because they "become not Princes, or great estates, nor them that write of their doings to vtter or report and intermingle with the graue and weightie matters". Her poetry then uses the "dark and unaccustomed" to address "grave and weighty matters", and frequently shifts between modes and codes within the course of a single work, the parts of a sequence and poems in the sections of her collection. But the book contains more than one piece suggesting you can have enough of etymology. Capildeo was briefly employed as a researcher for the ongoing revision of the *Oxford English Dictionary*'s some 231,000 entries (spelling, pronunciation, etymology, definitions and illustrative quotations). She addresses her subterranean working conditions in a poem called 'Sinking Lightwells', dedicated "For the OED (Etymology)":

> And what about those many operatives,
> those many office workers, well those people
> who clock in underground – can they be trusted
> to make good use of newly allocated,

all-natural, mind-expander, mood-lifter
no-excuse light?

Exploring the evolution of words might seem ideally suited to a poet, but she finds alienating both its working environment and the work itself. Far from stimulating a natural relation to the medium, it seems a Casaubon-like chimera in aspiring to a total knowledge of this now global language.

'Journal of Ordinary Days' begins by homing in on a verb aptly expressive of her frustrations:

Do I look like the sort of person who's not fit
to go out and buy a pen on her own? The phrase
'May I borrow a biro' is unspeakable
for its vocalic ugliness. The task in hand,
this third daze of work, is dis- and rearranging,
suspecting, assessing, keying in and tagging
all the historical spellings of the verb QUIT.
'That can't be a lot? QUIT is such a little verb?'
But people have been quitting for centuries, and
especially in Scotland, all in different ways.

As if "daze" were a variant spelling of "days", poets find their magpie inspirations in both cod-etymology and the real history that produces paradoxical words like the adverbial 'stand fast' and 'run fast' or the meanings of 'lie' so useful to jealous love posts. Yet the technical minutiae of etymology and the potentially endless task of sequencing variants is not an especially poetic prospect. Capildeo doesn't want to "borrow" a "biro" (a para-rhyme) because she needs to escape the office and "quit", so goes and buys one, not caring how long it takes. After attempting to get an "ethical biro" from a fair trade shop but failing, she falls into rap-like rhyming:

I look elsewhere, and not far off the ordinary
rewards the initiative: sell-it-all, old-fashioned,
like nineteen fifty-three, nearly customer-free,
a newsagent of the English variety.

This first part ends with a couplet made from an alexandrine and a pentameter: "Anything is better than going back to QUIT. / I can buy a pen on my own. I'm fit" – where the current "fit" in youth-speak, meaning *attractive*, consorts with more time-honoured senses. But as the poem implies, in the *OED*'s etymology section, she just don't fit and believes it's time to quit.

Capildeo's book addresses the vast hybridity of languages and an individual's interactions with their shifting terrains. "Sometimes", she writes in the poem's last section, "I dream in a language that is mine only in snatches", and explains how her sleep is populated by etymological shifts:

> Sometimes it is the actual people around me on a journey
> whose language drifts into another throughout my dreams,
> the prerequisite for transformation always being
> that both tunes already are familiar to my memory,
> so that the Irish have become Jamaican: the Spanish, Trinidadian...

The English-language 'subject' in speech whereby, grammatically, you are doing something, while, in experiments or the British state, you are the one to whom things are done, preserves agency within constraint. Thus the 'subjects' of verbs, the one qualifying the other (as in "John loves you" and "Mary loves you" – where John is your husband and Mary your mother), show how people are shaped by the words they speak, yet simultaneously shape the being shaped.

Lexicographers understandably make a fetish of words, their object of study, as when Herbert Coleridge, first editor of the *OED*, wrote that "every word should be made to tell its own story – the story of its birth and life, and in many cases of its death, and even occasionally of its resuscitation". Here lexicographers sound like interrogators with ways of making words talk: but words don't have lives and deaths, and citations in dictionaries are the tips of titanic usage icebergs. Words are no more nor less than attempted transcriptions of noises made by people with their mouths and, as with "quoof" in Paul Muldoon's family or "glory" to Humpty Dumpty, what we can mean by our oral and aural inventiveness is limitless. Capildeo's poem has it that the poet is the person who goes to buy the writing implement, not the one who tags sampled, and inevitably partial, usages. Nevertheless, it's good to know that "divan" may be related to the Persian *debir* meaning 'writer' – William Cowper, for instance, who began his long poem *The Task* with "I sing the Sofa" to "seek repose upon an humbler theme".

Aboot The Labour:
The Possibilities Of Scots

W.N. HERBERT

The exciting emergence in Carcanet's *New Poetries V* of William Letford, a poet writing in both English and Scots, is an opportunity to re-examine the energies and tensions many Scottish writers experience between their literary languages.

Not simply independent of English, literary Scots begins in a specific local dialect like Glaswegian, extends back toward the 'Inglis' of the late medieval makars, and forward through conventional English to the diversities of world Englishes as spoken, written, sung and broadcast in the US, India, Africa and Australia.

Our privileged position as the first 'other' to English's norm, plus our extensive literature, means that Scottish writers' decisions about which Englishes we use can be as complex as our decisions, both inside and outside Scotland, as to how we speak.

Nicholas Lezard's enthusiastic review of Letford's work in *The Guardian*, though not primarily focused on his use of Scots, is engaged by his linguistic vigour in a way that opens this area for discussion. He praises the use of playful Morganic layout in 'Moths', professing himself delighted that "its first proper line [consists of] the words 'fucking moths'."

He then quotes from the poem which makes the most interesting use of Scots, 'It's aboot the labour', adding in parenthesis: "(Letford... has a real job – as a roofer – and his poem about getting up to work in winter when it's still dark should make poncey southern metropolitan softies like me a little bit ashamed of themselves.)"

Lezard's generosity is to be welcomed. But his response suggests there are two ways of looking at this issue. On the one hand, Letford conveys a sense across these poems of the deep consonance between language and action, of what it is like to be engaged in craft – to be positioned in the world most forcefully through the act of making – and he creates a dynamic interrelation between the making of the poem and the act of labour it describes, making one event of two activities.

The other is the suggestion (suspicion, fear...) that Scots is thrilling and, somehow, more 'real', because it is definable as working class. In this case, any further engagement with either craft or Scottishness may prove surplus to requirements. But of course the point about these poems is precisely that Letford moves between his languages, thereby suggesting that facticity isn't simply a matter of saying "fuck".

There is a further layer of referencing here. These two men, wielding hammers and nails in an elevated position and discussing culture in phonetic Glaswegian, clearly recall Tom Leonard's 'The Good Thief' in which Catholic /Protestant bigotry is replayed upon the cross, as the eponymous speaker addresses Christ: "heh jimmy / ma right insane yirra pape / ma right insane yirwanny us jimmy / see it nyir eyes / wanny us". (Note that wicked homophone 'insane' for 'in saying'.) Letford's poem turns on a similar moment of questioning:

> heh Casey did ah tell ye a goat
> a couple a poems published
> widizthatmean
> widayyemean
> dizthatmeanyegetanymoneyfurrit
>
> eh   naw
> aw   right

Of the two crafts on display here, only one is remunerative, and Casey's question conflates two supposedly distinct economies: paid physical labour and the purportedly unassayable labour of the mind – both are revealed as modes of exploitation.

Leonard's use of Glaswegian was part of a broader critique of our assumptions about people based on their speech, and his writing also scrutinises the registers of politicians, religious figures and academics. As he says in his introduction to the anthology *Radical Renfrew*, "To understand Literature is to understand a code, and the teacher is the person trained to possess the code that Literature is in".

Moreover, it was a reaction to the hieratic Scots vocabulary and metrics of a preceding generation as exemplified by Hugh MacDiarmid, whose Scots might contemplate 'The Innumerable Christ' or address a 'Hymn to Lenin', but who approached the working class through sermons and lectures ("Look, Wullie, here is [Lenin's] secret noo / In a way I can share it wi' you").

There is a further layer of irony to Leonard's work, hinted at in the "impossible" exam questions of the later 'Four Conceptual Poems': just as literature is a code, so too his use of Glaswegian is an encoding gesture, only opened for some through acts of cultural interpretation such as those of academics and reviewers. This points to the directedness of Scots, the way in which it appears to be for Scots firstly, and only open to "southern metropolitan softies" (and others) as an act of over-hearing.

Of course lyric poetry itself, its 'I' addressing an intimate 'you', is already a kind of overhearing for the reader, but there is a danger that in this game of British whispers the plurality of the Scottish linguistic sensibility gets overlooked in favour of shorter, sharper impacts. We smile when Lezard writes of another of Letford's poems that it's "the most thoughtful poem about getting head-butted that I have ever read" – but we also wince.

I'm reminded here of the work of Alison Kermack (now Flett), published in the first edition of Donny O'Rourke's influential anthology, *Dream State*, in 1994. What O'Rourke called her "scathing urban Scots" was certainly one of the spikier examples of Leonard's influence. However, O'Rourke omitted her from the second edition, saying sadly, "she hasn't been writing". In her 2004 book *Whit Lassyz Ur Inty*, we discovered she had indeed been writing – only in a gentler mode, reminiscent of Ian Hamilton Finlay:

> an we stood taygither
> at the endy thi peer
> lookn outwards
> trine tay mind oan
> thi namesy thi boats
>
> cum hame boy gordon
> we called oot
> cum hame girl mina
> john L...

Perhaps to limit Scots to literary head-butts is to direct it into a prematurely dead end. It's still too early to assess the full potential of William Letford's work, but it's possible to conjecture on the evidence of *New Poetries V* that another intriguing restatement of the possibilities of Scots, and of the subtler relationships between that language and English, may be underway.

℘

English Intelligences: R.F. Langley's Late Poems

JEREMY NOEL-TOD

A recent small-press anthology, *Certain Prose of The English Intelligencer* (Mountain Press, 2012), valuably collects some early exchanges between the contemporary poets often referred to as the 'Cambridge School', after the university town that connects them. It's a book for anyone interested in an alternative to the Hughes-Larkin prizefight of post-war literary history. Prominent in the urgent pages of this "worksheet", circulated between 1966 and 1968, were Andrew Crozier, Peter Riley, Barry MacSweeney and J.H. Prynne.

The name of R.F. Langley appears only once, on a list of recipients, alongside his West Midlands address. He was not a contributor. Thirty years later, however, on his sixtieth birthday, Langley was honoured by a poetic *festschrift*, *Sneak's Noise* (1998), which included tributes from Prynne, Riley and MacSweeney, as well as another *Intelligencer* reader in the Midlands, Roy Fisher. Like Basil Bunting and W.S. Graham before him, Langley was a patiently experimental poet whose modest, first-water oeuvre united admirers.

The following year, he retired from his career as an English teacher and moved to Suffolk. The publication of a *Collected Poems* (Carcanet) in 2000 – collecting seventeen pieces – marked the start of a late flourishing that eventually left nearly fifty poems. A second volume, *The Face of It* (Carcanet), followed in 2007, and Langley continued to appear in periodicals until his death in early 2011. An uncollected later lyric, 'To a Nightingale' (2010), went on to win the Forward Prize for Best Single Poem that year.

Like many Langley poems, 'To a Nightingale' locates its speaker in a rural English landscape, focusing on the small details of the scene with an almost hallucinatory loss of identity:

> Nothing along the road. But
> petals, maybe. Pink behind
> and white inside.

This is recognisably the same poet who saw, in the first poem he preserved, the "white lining" of willow warblers' wings, "glint[ing] like signals". That poem, 'Matthew Glover', shows the transatlantic influence of Charles Olson, whose ambitions for poetry as a project of intensively local research ("dig one thing") are repeatedly echoed by the essays of *The English Intelligencer*, which include a philological investigation by Prynne of the runic sign 'wynn', and an extended speculation by Riley on prehistoric burial customs.

A fascination with the archaeological layering of culture recurs in Langley's work. 'In the Bowels of the Lower Cave' (2008) vividly realises T.S. Eliot's fancy – in 'Tradition and the Individual Talent' – that the modern poet must inhabit a collective "mind" reaching all the way back to "the rock drawing of the Magdalenian draftsmen". The first half of the poem tracks the twentieth-century discovery of animal paintings and engravings at Gargas in France back to an imagined intimacy with the Paleolithic artist making the marks:

> ...to reconstrue
> outward from the nip of
> shadow pecking at the graver's tip

(Prynne and Langley, who met as students at Cambridge, visited the caves at Lascaux before they were closed in 1963).

The second half of the poem cuts from the "this, that and thus" of Langley's verse-lecture (he also gave extra-curricular Art History classes) to a passage of diary-like immediacy, as the poet's pen continues to make its twenty-first-century marks on paper: "Eight thirty. Tenth February. Frost smokes." The speaker's precise confusion of "water clucking" in a ditch with distant hooves on tarmac leads to a conclusion where the imaginary and real entwine – as so often in this writing – in an impossible, welcomed fullness of knowledge:

> Nothing comes trotting round the corner but
>
> he is close alongside nonetheless, in
> optimum position, the small bearded
> horse at Niaux. I see him as he is.

Langley's acceptance of life as a collage of accuracy and chance was praised by Peter Riley in one of the first critical appreciations of his work, 'A Poetry in Favour of the World' (1997). Elsewhere, Riley characterised it as a "distinctly challenging" poetry able nevertheless to remain "remarkably gentle in its craft of verbal precision". Langley practised close observation throughout his life in the extraordinary prose of his private journals. A selection of entries was published by Shearsman in 2006, and a number were reworked into poems. 'At South Elmham Minster' (2010), for instance, began life as the record of a visit to the Suffolk ruin in 2004. The journal's simple closing comment ("today I like very much this all that there is") inspires a reflection in verse on how the archaic word "thiskin" ("of this kind") chimes with the specific name of a thing to hand: "thistles".

After 'Matthew Glover', Langley moved away from Olson's ideal in 'Projective Verse' (1950) of an 'open field' poetic towards a formally compressed prosody of his own, tightly hedged by rhyme and line. He remained committed, though, to Olson's micro-principle that the syllable "is the king and pin of versification, what rules and holds together the lines". Here, for instance, is Langley's wonderfully modulated description of the wake of a boat on water-stairs in Venice: "swells sledge clean sheets off the treads, / and swill their slops on the risers" ('Di Fronte', 2010).

Langley's most celebrated sequence, the *Jack* poems (1998), conjure a figure of the imagination from all the definitions under that word in the *Oxford English Dictionary*. The Lancastrian pastoral poet, Michael Haslam – whose reflexive, rhyming, wildly iambic style Langley greatly admired – greeted them as "an English Classic" in his contribution to *Sneak's Noise*: "Roger Langley called it Jack. / And we knew it was that" ('Party Spirits').

Last summer, the *Sneak's Noise* poets gathered again for a memorial reading alongside other similarly single-minded practitioners: Denise Riley, Tony Lopez, Thomas A. Clark. Haslam reprised the poem, with an elegiac addendum:

> How could a heart like Roger's fail
> with such a knave as Jack to set the pace?
> The case is grave, and yet it's not too glib to state
> that through observant wit, throughout the poetry
> of R.F. Langley, the spirit lives.

℘

All the uncollected poems referred to in this article can be found in back issues of *PN Review* and *London Review of Books*.

Denise Riley:
Writing Our Difficulties

EMILY CRITCHLEY

The constellation of writers and thinkers surrounding Denise Riley's work is as various as it is starry. It includes poets R. F. Langley, Douglas Oliver, John James and others from what's known as the 'Cambridge school' – including an especially important friendship with Wendy Mulford, who we have to thank for the fact that Riley began to publish when she did (in the 1970s) and possibly at all. It also includes North Americans Lisa Robertson and Fanny Howe, and the New York School writers (similarly interested in painting) Frank O'Hara, Ted Berrigan, James Schuyler, Alice Notley, early John Ashbery and Anne Waldman. And then Apollinaire, and also early Auden, Herrick, Marvell, Blake, Wordsworth and 'The Child Ballads'. The constellation could be mistaken for a universe depending on how far back your eyes can go.

Interestingly, Riley puts many of the Renaissance and Romantic influences – along with sixties pop lyrics, folk, and Schubert's lieder – in the category of 'song', and even a cursory glance at Riley's poetry reveals the importance to her work of lyric-as-music, "an obstinate attachment to [...] some musical brightness" as she put it 'in conversation' with Romana Huk (*PN Review*, 1995). Often, however, this bright music proves the vehicle for some of Riley's most difficult intellectual and emotional work, as evidenced by her most recent published poem, 'A Part Song'. A sequence of twenty short poems, about the death of her adult son four years ago, 'A Part Song' appeared in the *London Review of Books* this February after a long gap in which Riley no longer seemed to be publishing poetry. In its varied metrical virtuosity, we hear heart-breakingly preserved some of "the piercing violence of the early ballads" ('In Conversation') – all the more affecting for the brightness and, at times, mock humour of the poem's rhymes, cadences and other musical effects (such as alliteration):

> Oh my dead son you daft bugger
> This is one glum mum. Come home I tell you
> And end this tasteless melodrama   ('A Part Song, vii')

Always in Riley's poetry there is this tension between the relief offered by its lyrical, often breezy surfaces – "a thin and fragile layer of cadence" ('In Conversation'), made up of popular, 'misremembered [song] lyrics' (the title of a poem), and "exaggeratedly lush" descriptions of nature or paintings, for instance – and the serious thought-work the poetry does on identity and gender: 'sex' as Riley prefers to call it, politics and interpellation, and the role or rather the "force of language" in all this. Riley's important philosophical writings (*The Force of Language* is one of these) explicate much of this work far more rigorously than the poetry, of course, but her poems get to test that work out musically and (thus) feelingly, to be "illustrations of a linguistic emotionality at work" (Riley, *The Words of Selves: Identification, Solidarity, Irony*). Indeed, she is a poet who manages to underline complex emotion as an index of thought *per se*: "It is called feeling but is its real name thought?" ('A Shortened Set').

We see this especially in the poems' constant experimenting with plural voices – a range of subjectivities which are never, in the theatre of writing, exactly the poet's own anyway, but which Riley theatricalises more than most, trying on different identities – "in case they offer comfort", as Peter Riley recently put it in *Fortnightly Review* – as her poetry does styles, colours and materials (all three abound as motifs in her work). That she does so so successfully is precisely what makes her poetry feminist: not in the vacuous, glossy magazine sense of presenting women's choices qua as many different lipstick colours as the free market can sell, but, profoundly, by refusing to be hemmed in or pinned down (puns intended) by sex. "My aim," she writes in *"Am I that name?": Feminism and the Category of 'Women' in History*, "is to emphasise that inherent shakiness of the designation 'women' which exists prior to both its revolutionary and conservative deployments. [...] The cautionary point of this emphasis is far from being anti-feminist. [...] A political movement possessed of reflexivity and an ironic spirit would be formidable indeed".

The constant mobility of Riley's personal pronouns, from "I" to "you" to "she" to "she-husband" (in 'A Note on Sex and "the Reclaiming of Language"', *NB* the arch scare quotes around that phrase) to "not-me" ('Affections must not'), and so on, is part of that reflexivity and that ironic spirit – both so important to Riley's work, as is the personal/anti-personal/personal directness (almost in the spirit of Frank O'Hara's 'Personism') of those voices, which makes up not so much the guarantee of authorial authenticity, as its very testing ground. As if Riley were querying the artificiality of attempts to be personal in language, even, sometimes the point of poetry altogether (the vanity behind wanting to be listened to is almost too repellent to her):

This
representing yourself, desperate to get it right,
as if you could, is that the aim of the writing?
[...]
    Are you alright I ask out there
Straining into the dusk to hear.
I think its listening particles of air
at you like shot.
You're being called across your work
or – No I don't want that thought.
Nor want to get this noise to the point
it interests me. It's to you. Stop.   ('A Shortened Set')

This calling between and spilling across subjectivities for me comes close to the power, so much admired by Riley, of colour in abstract painting "to spill off its own margins and break the separate stillness" or solipsism of the painter ('In Conversation'). In 'Denise Riley *passim*' Robin Purves connects Riley's literary efforts not to be pinned down by "biographical singularity" with her interest, in several of her poems, in the "idea of painting as a site of freely minimal subjectivity", their "description of virtually autonomous behaviour on the part of pigments [of colour] that move under their own steam". Moreover, Purves associates both with Riley's flexible, philosophical thought – especially as laid out in *The Words of Selves: Identification, Solidarity, Irony*. In this Riley claims irony as an invaluable way both to maintain and to check "categories of social being" – such as "writer" or "woman" – which are only ever provisional, "arbitrary", and which must be so "for there to be mobility, and life, within political thought".

A mobile "I" or "you": not the kind of delimited, centreless flux that has become fashionable in postmodern culture, but a social subject capable of adapting and responding to each new circumstance: "ultraflexible" like the "hope" in 'A Part Song': "You principle of song, what are you for now / [...] Slim as a whippy wire / Shall be your hope, and ultraflexible". For Riley does remain, by her own admission, "naively hopeful" (*The Words of Selves*) about language's capacity to undo enforced categorisations – through ironic iterations and other means. And despite the difficulties and doubts registered by and in her verse – "Neither my note nor my critique of it / Will save us one iota. I know it. And." ('A Part Song') – that "And" remains, as does her impulse to keep noting it all. For that we can only be glad.

Works referenced
Poems from *Denise Riley: Selected Poems* (Reality Street Editions, 2000)
'A Part Song', *London Review of Books*, Vol. 34 no. 4 (9 February 2012)

Polaroidy: The Revels Of Geraldine Monk

ADAM PIETTE

Geraldine Monk, *Lobe Scarps & Finials*, Leafe Press,
£8.95, ISBN 9780956191946

Geraldine Monk's poetry is always a cause for celebration: its wit, wily, word-enthralling, is an education into regions of double-jointed nonsense and joy, sensualities of syntax, rapturous vision, satirical intelligence facing down iniquities, a multivocalised performance of ulterior identities, in a carnival zone, various, gracious, outrageous. To write wittily is usually taken to mean something trivialising: what depths of shallowness in the idea of 'comic verse'. Monks's poetry redefines the comedic, beyond region (though including a strong form of northern blunt clairvoyance and word-savouring), beyond mere slapstick turn or innuendo (though the poems glory in music-hall spiritedness and catch-song chorus effects), beyond the punchline or the routine (though she is expert at the deep pun, the savage turn and the many masks of theatre). It unleashes the energy of wit into the fieldforce of language itself, wit as an improvisatory method crafting voices through electrically alive craftiness. *Lobe Scarps & Finials* may not have the tremendous ambition of the great work on the Pendle witches, *Interregnum*, nor does it meditate so imaginatively on the dead as with the extraordinary work on Mary Queen of Scots in Sheffield Castle, *Escafeld Hangings*. But it is seriously witty and as a collection invents a variety of comedic sequences that explore art and being in the world in ways that are absolutely Monk-imprinted through and through.

We begin with a joyous *jeu d'esprit*, the sequence 'Glow in the Darklunar Calendar', which is a cross between a strange monthly diary of ponderings of personal experience in the year 2010, and a tongue-in-cheek, seriously performed compendium of names for the full moon taken from many cultures. The effect is uplifting, for there on the page merge a formal moon chant with the caustic, odd and curious Monk observations, polaroidy somehow yet inviting, companionable. Two wits fuse: the wit that takes from the world of language forms and rewrites them (so the moon poem, made bankrupt by Apollo 11, is rewomanised); and the wit which is a friend's jokiness, poking at the world as it happens, making something of the sparks that fly. So we get the views on a funeral, a trip to the States, a women's poetry event, as well as on the banking crisis, a shooting, global warming etc., but the double wit lifts it from the dailiness and the 2010-ness into a cross-hatched community of events: it *makes a sequence from them* – offered, as it were, to the goddess of poetry herself. And what beauty to the lines as a result of the variousness of the wit!

> volcanic ash paves a fright of
> sheer craftless skyways –
> birds wing their atavistic myth
> fresh of flesh & wimberry-eyed
>
> > *Wild Cat Moon*
>
> Mind the celestial gap.

Eyjafjallajökull is the occasion, and the future is served by a droll meditation on that current affair, in April 2010, when an ash cloud grounded planes and the there was wonder at the contrail-free skies, the sudden green thought that a non-technological earth might be more than remembered. The dream of flight without airplanes becomes a sensing of poetry-making as wit, though, when Monk puns on 'craft'. If her poetry is so 'craftless' (as woman's poetry, comic poetry and non-mainstream writing is often held to be), then where did that superb line "fresh of flesh & wimberry-eyed" come from? The *words* are fresh, have flesh of a kind, and tap into the local ("wimberry"), recall Hopkins, Berryman, Dylan Thomas, Dickinson with the colour, the natural detail, the ampersand even, yet are entirely Monk-like. They sing, and call out both to the sweetly imagined cat moon of ancient folk poetries, and to the free-wheeling jester within language, and her way and play with words.

The collection then moves to its second sequence, *Print & Pin*, and conjures up a hilarious, moving thought experiment based on newspaper lost and found columns, except here what are lost and found are hybrid poem creatures, soul animates, riddle-me-ree objective correlatives with bells, lights and whistles. Each poem, centre justified, short-lined, gestures towards concrete poetry, inviting readers to think about word shapes, the forms we give thoughts, what happens when we allow words to take over and play. One of the FOUND poems imagines "a flock of half-baked / l-o-n-g-t-h-o-u-g-h-t-s / colour of uncooked ham-hock / slightly needy", and asks us to ring and claim. It is a delightful improvisation on poetry as emotional appeal, yet trumps that too with the zaniness, the crazy loving going on, the zest of it all.

If the opening sequences invite and delight, the third takes us down a quite different sequential route: it stakes itself out as a 'Collaboration with John Donne', recasting the words of 'A Nocturnall Upon S. Lucies Day', that great meditation on time and eternity. It forms part, therefore, of Monk's extraordinary imbrication of her own vocal repertoire with sixteenth- and seventeenth-century writing and history. As Monk has said herself of these collaborations with the dead, they allow her to counter the "self-policing of the mind" through "the invasive undermining and enhancement of an other", whilst not relinquishing control over her own work. To "invite Donne" to jointly co-author a dismantling of his own work is outrageous perhaps, but the effect is Donne-like at another level:

> the body must
>   be
> here body
>   be
> must
>   be
> here
>   be
> body must
>   be
> the body must
>   be
>
>   as shadow

This has the quality of Donne's sermons, whilst also being truly Monk-like in the pitch of its feeling through of questions of being, of the need to secularise and live as bodies here on earth, to *be* as bodies, yet also as shadows, shadows on a page. It is quite wonderfully done, and wonderfully Donne.

The other sequences here are as good. The superb 'Poppyheads' imagines a persona floating round her garden pond, performing the variety of ways garden imagery informs and constrains culture, in high parody: "One serpent short of Eden / the garden warbles in excelsior. Weeding / pesky Latin names I grow bonnets / slippers and ladies ear-drops. Hock-hock." Never has the gardening mania of England, the Edenic control of sweet ladies and the flowery rhetoric of fake poetry received such a good kicking. 'Racoon' uses comedy to riff on America, on the animals not seen, on a dream of English poets as useless phoenixes (thus parodying American prejudice), all in the generous, lively and loving spirit of a good O'Hara poem. All in all, this collection is simply luscious, and a quite marvellous display of the virtues of wit and sequence, now and in the world.

ℬ

The Taste Of TEN: Questions Of Propriety

DALJIT NAGRA

*T*EN is an anthology that supports new black and Asian talent. The new poets are all working towards a first collection or have yet to publish a collection with a 'big' publisher. The anthology may be viewed as a calling card to publishers, with the funding body, Arts Council England, in effect saying to publishers: "We have helped to nurture the talent of these new poets and have tried to bring their talent a bit more in line with your 'taste'. Now would you be prepared to take a risk by publishing some of them, please?"

I use the word 'taste' because that is the word used by Bernardine Evaristo in her brilliant and stirring introduction to the anthology, "What if poetry publishers, nearly all of whom are white and male, used their position of power and privilege to be more proactive in actively seeking out new voices away from the usual networks? It might mean publishing beyond personal taste."

This white male dominance within poetry publishing can be seen to reinforce a certain 'taste': this community's taste reflects its value system as well as its cultural aesthetic. The problem arising from the homogeneity of publishers' taste means that the breadth of experience lived in multicultural Britain is not explored. *TEN* challenges these publishers to respond by broadening their taste in response to this cultural shift. The new 'taste' may unsettle that of the dominant community because it represents a different aesthetic and observes a different code of decorum. In the following I will think through one example of the new 'taste' by considering two practical examples and explore how such provocation is healthy for readers and writers of poetry.

My editing skills are have been honed within the dominant traditions of British poetry, so are also based on a 'less is more' sensibility. This strategy regards subtlety of sensibility as a mark of good taste. Some poems in *TEN* do not conform to this aesthetic, however we will see that it is unfair to judge such poems by these standards. As co-editor of *TEN* I was challenged by several poems that I initially found too direct in style and content. By stepping back from my conservatism I was able to appreciate this different taste.

Malika Booker's 'Pepper Sauce' is a powerful poem about a grandmother's brutality. The grandmother prepares a burning hot pepper sauce concoction and "she pack it deep in she granddaughter pussy" whilst the latter is tied to bedposts. The narrator then lists some of the ways in which the girl is affected by the sauce. I advised Malika that she should cut back her list as it overstated the case:

> I hear there was one piece of screaming in the house that day, Anne bawl
> till she turn hoarse, bawl
> till the hair on she neighbour skin raise up, bawl
> till she start to hiss through her teeth, bawl
> till she mouth make no more sound…

I realised later that my judgement may have been based on a taste which lacks sufficient insight into the experience of the speaker and the poet. From her aesthetic stance, Malika felt it necessary to fulfil the graphic demands of the narrative.

Another poem is Mir Mahfuz Ali's 'My Salma' which is set in Bangladesh during the civil war. Mahfuz describes a gang of soldiers making the speaker attack his own lover, Salma, and the soldiers subsequent gang-rape of her: a soldier "laughed as he pumped / his rifle-blue buttocks in the Hemonti sun." The contrast of occasionally pleasant language hounded by violent detail would offer a powerful enough ending, but Mahfuz takes the poem a stage further. In the editing stage I advised against this. Once again, however, my editing sensibility felt inadequate. The speaker progresses to tell us, and I feel uncomfortable saying this, that the soldiers then mutilated Salma, and the speaker comes across a dog who is eating Salma's headless body. This remarkable poem does not end there, and my cosy middle-class sensibility forbids me from paraphrasing further! I'll quote the final lines instead:

> Then I saw against the deepening sky
> a thin mangey bitch, tearing at a body with no head,
>
> breasts cut off in a fine lament,
> I knew then who she was, and kicked
>
> the bitch in the ribs, the same way
> that I had been booted in the chest.

I suspect both poems may not go down well, in terms of taste, with publishers. Both Malika and Mahfuz feel it important to challenge decorum, to go that bit further so the violated and the violator are properly evidenced. From my Punjabi background I hold some sympathy with the need for strong evidence as more respectful to a disturbing incident than a coded response. In harsh circumstances, subtly of language is distrusted, even photographic evidence is preferable.

I don't know if Malika and Mahfuz share this same distrust of subtle language in the face of disturbing scenarios, all I know is that I need to reach out and appreciate a sensibility that is different to the one I was taught to respect when I studied English at university here in Britain. The poets in *TEN* are all highly educated and skilled practitioners of poetry and they have helped to broaden my taste. I hope publishers will be prepared to broaden the poetic landscape. As Bernardine says in the Introduction, "What if the editors consider the idea of representation not as an imposition to be resisted and rejected, but as a demonstration of inclusivity?"

B

POETRY REVIEW SUMMER LAUNCH **AT POETRY PARNASSUS**

Mapping the Delta

Readings from **A.B. Jackson**, **Lorraine Mariner**, **Geraldine Monk**, **Denise Saul**, introduced by *Poetry Review* guest editor **George Szirtes**

FREE EVENT. ALL WELCOME
Saturday 30 June, 2pm

The Clore Ballroom, Royal Festival Hall
Southbank Centre, Belvedere Road
London SE1 8XX

A.B. Jackson photo: Susan B. Breakenridge
Denise Saul photo: Naomi Woddis
George Szirtes photo: Clarissa Upchurch

For more information about our events and the summer issue of *Poetry Review*, visit www.poetrysociety.org.uk

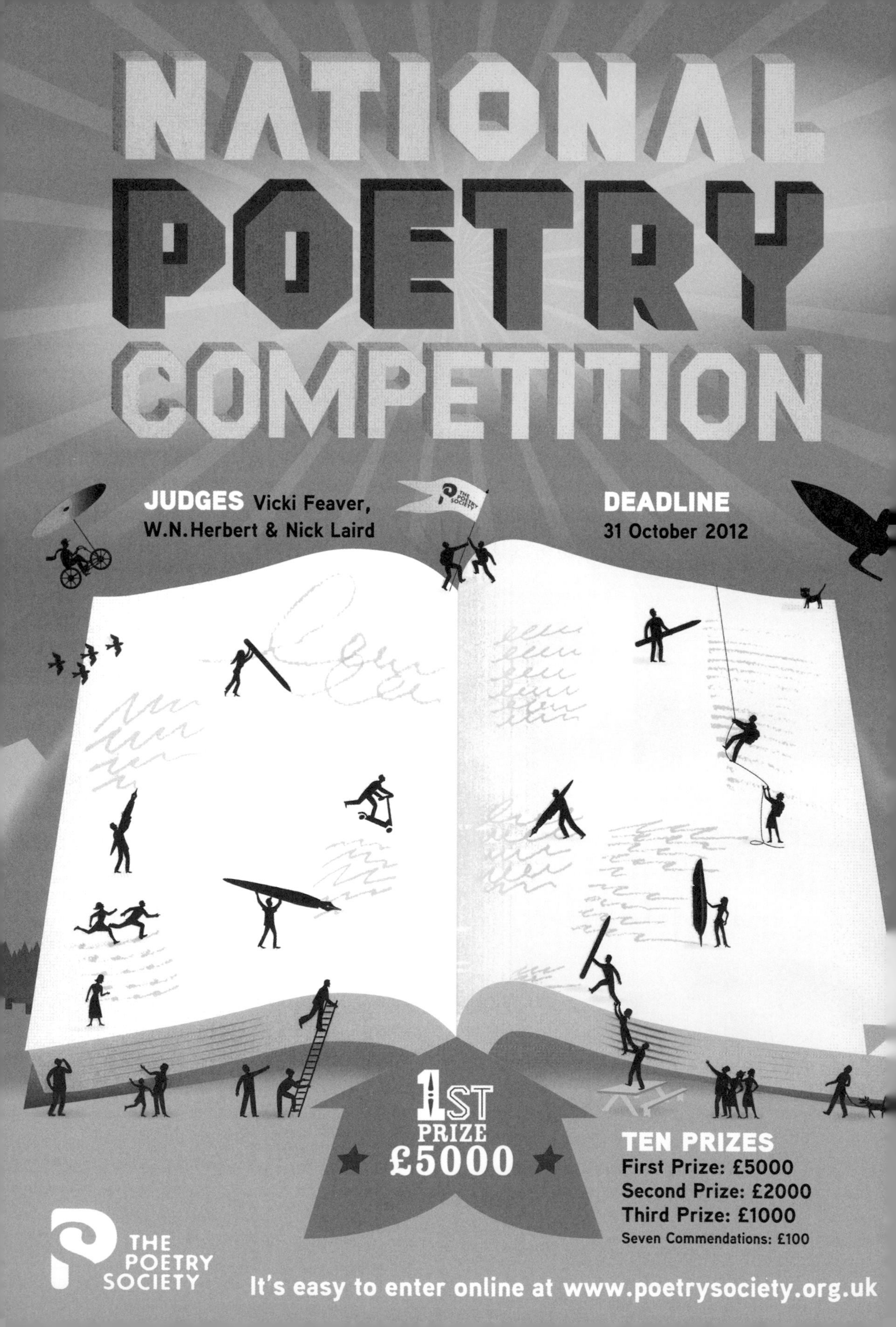
NATIONAL
POETRY
COMPETITION
JUDGES Vicki Feaver,
W.N.Herbert & Nick Laird
THE POETRY SOCIETY
DEADLINE
31 October 2012
1ST
PRIZE
£5000
TEN PRIZES
First Prize: £5000
Second Prize: £2000
Third Prize: £1000
Seven Commendations: £100
THE POETRY SOCIETY
It's easy to enter online at www.poetrysociety.org.uk

REVIEWS
&
ENDPAPERS

ॐ

... that the world of particular creaturely
things is a world of speaking likenesses...

– Peter Scupham on Clive Wilmer

Circumambient Weathers

PETER SCUPHAM

Elizabeth Jennings, ed. Emma Mason, *The Collected Poems*, Carcanet,
£25, ISBN 9781847770684;
Clive Wilmer, *New and Collected Poems*, Carcanet, £18.95,
ISBN 9781847770523;
Keith Chandler, *The English Civil War Part 2*, Peterloo, £7.95*

Ramshackle Elizabeth Jennings is tidied here into one thousand and twenty pages of lyrics closely tripping on each other's heels, with a Preface, an engaging photograph of a laughing girl not yet in the thralls of the Wicked Witch of Poetry, and a substantial Afterword. She was born in 1926 and died in 2001, and, as she came to maturity, poetry was, as usual, divided between those who believed poetry shadowed out in words more mysterious truths than the words themselves provided – the Neo-Romantics, Treece, Gascoyne, Thomas in one camp, and those in another who described with an ironic intelligence the cage in which we usually find ourselves – the Movement, itself in some ways deriving from the Amis / Ewart world of Beer and Bints, the wartime Officer Class whooping it up in the Sergeants' Mess. The third way in poetry was the way of faith, specifically the Christian faith and assorted symbolisms, with Eliot as the Master – Kathleen Raine, David Jones, Anne Ridler and Elizabeth Jennings, herself a Roman Catholic and a troubled one, for whom all gingerbread came with guilt. I remember my first Jennings, 'The Counterpart', cut out of *The Listener* and scrapbooked in the late 1950s; though early, it has many of the qualities that I admire in her work: the formal elegance and a strong sense, as Yeats put it, of the trembling of the veil, the parting of curtains between the visible and invisible worlds. It also has one of the flaws attendant on the symbolic tradition, a reliance on heraldic imagery and a fondness for abstract discourse:

> The symbol on the shield, the dove, the lion
> > Fixed in a stillness where the darkness folds
> In pleated curtains, nothing disarranged:
> > And only then the eye begins to see.

---

*Now available from www.fairacrepress.co.uk or Amazon, £4.95, ISBN 9781904324522

Well, yes but no, and though a hieratic approach is a constant in Jennings's work, there is plenty, too, which gives us a fuller and sometimes a more colloquial reading of the intermeshed worlds she is at home in. The sometimes over-careful careful arrangement of experience is a prophylactic against chaos, a healing remedy against the background of a long life survived against a background of physical frailty, mental disturbance, attempted suicide, and a nature attuned to loss and elegy:

> Even when grief is threatening, even when hope
> Seems as far as the furthest star.
> Poetry uses me, I am its willing scope
> And proud practitioner.

Her lucid, questioning and puzzled poems develop freedoms, draw closer to that unhieratic natural world she also lovingly celebrated, as in 'The Feel of Things', where she praises

> The feel of things, the nap, the fringe, the sheen,
> To catch the light that circles round the cat
> Brushing him as you groom a horse.   To show
> The rueful tumble of a head of hair,
> Its curl, its fall.

That poem closes with a physicality which has come a longish way from my remembered 'The Counterpart' of half a century ago: "The poem is a way of making love / Which all can share. Poets guide the lips, the hand". This book is a treasure-house.

Clive Wilmer's *New and Collected Poems* are consonant with that of Elizabeth Jennings. 'The Goldsmith', where an amulet is engraved to make "A little space of order: where I find, / Suffused with light, a dwelling for the mind" has its counterpart in 'The Counterpart'. Though he has not taken the rigours and doctrines of one particular faith as a given, Wilmer's steady, exact and meditative work is lit by the constant belief that the world of particular creaturely things is a world of speaking likenesses, that, as Christopher Smart put it in his 'Nativity Ode', God "is incarnate, and a native / Of the very world he made."

His early book, *The Dwelling Place*, has as epigraph a wonderful quotation from Wilmer's admired Ruskin, in which Ruskin talks of a man's religion as being a "dwelling-place" built from custom, region, family pietas and the accidents of personal living. As Ruskin did, he particularises lovingly, as in 'Minerals from the Collection of John Ruskin':

> It was the Lord's design he made apparent
> These bands and blocks of azure, umber, gilt,
> Set in their flexing contours...

It comes naturally to Wilmer to praise, to give, to admire, to dedicate poems, to offer tribute, whether to George Herbert or an old schoolmaster; the book is full, too, of those "flexing contours" of the world, the thinginess of things. He is with Elizabeth Jennings in his constant celebration of life, the play of the living powers in hedgerow, meadow, the circumambient weathers. This is a fine collection, serious, loving and considered.

And now, with a fanfare, comes the Court Jester, Keith Chandler, though as good jesters do, he speaks unwelcome truths. The title-poem is a hilarious futuristic re-run of the English Civil War, with Harry dashing about with Challenger tanks doing his Prince Rupert Act, King Charles exiled to South Georgia and the New Levellers (temporarily?) in charge. His tone is usually jaunty-demotic, but also mock-jaunty; the most serious sequence is a chilly set of poems in different personae, 'Postcards from Auschwitz'. Camp Doctor, Kapo, Bureaucrat and others offer their meaningless justifications, and though Kingsley Amis once wrote "In love and war / Despatches from the front are all", the unwriting dead can send no despatches we do not write for them. Keith Chandler's book is enormously readable, closing with two memorable longer poems, one on the East Coast floods of 1953, and a witty piece in the persona of escapologist Harry Houdini, watching fans try their yearly evocation of his spirit, though his 'ghost' suffers ghostly deconstruction:

> I never spoke.
> Someone else, a Limey puts these thoughts
> Into my mouth, aims to ventriloquise
> my voice, to ghost-write my non-ghost.
> I repeat. There is no come-back. I am dead.
> I am not hovering somewhere overhead.

Read this hilarious and mordant book, and consider what Elizabeth Jennings and Clive Wilmer would make, philosophically, of that last quotation.

Peter Scupham's *Collected Poems* appeared in 2002. His most recent collection is *Borrowed Landscapes* (Carcanet, 2011).

Play More Softly

TARA BERGIN

Astrid Alben, *Ai! Ai! Pianissimo*, Arc, £8.99, ISBN 9781906570729;
Gregory Woods, *An Ordinary Dog*, Carcanet, £9.95, ISBN 9781847770783;
Adam Thorpe, *Voluntary*, Cape, £10, ISBN 9780224094177;
Paul Durcan, *Praise in Which I Live and Move and Have my Being*,
Harvill Secker, £12, ISBN 9781846556272

Intrinsic to each of these four new collections of poetry is the notion of privacy. In each case, the strongest and most touching poems are the ones which appear to come out of an intense, personal experience – sex, love, mourning – the point at which intimacy becomes formalised, yet when the formal is powered by the intensity of feeling.

The title of Astrid Alben's first full collection, *Ai! Ai! Pianissimo*, is intriguing, yet slightly misleading. Alben's poetic strengths are arguably best displayed in the poems which engage chaotically and abruptly with their themes, rather than those which "play more softly". Her depiction of the violating experience of relationships, for example, makes impressive use of unusual imagery and irregular syntax and rhythm. In 'The View,' love begins – perhaps – with a man spilling tea in a woman's lap and then fetching a cloth "to wipe her clean", while in 'Bucking Bronco', "he bucks her outwits her outdoes her". Similarly, in 'Take-away Heart,' the distance between "he and she" is weirdly, yet clearly, conveyed: "Let's get take-away she says. / *I need to be in your hair*". This collection is full of interesting viewpoints and narratives: in 'To the Highest Bidder,' a couple attend the auction of their own failed relationship, where everything, "including the space across the table at which they met," goes under the hammer. Here, Alben's strange grammar and sparseness of punctuation lends her poetic language a disturbed and disturbing quality, one that could certainly be further explored in subsequent collections.

Gregory Woods's new collection, *An Ordinary Dog*, contains countless examples of his ability to handle material with a compelling combination of control and energy. His overt celebration of the physical body is matched by a delight in rhyme and form. Sometimes this can seem like extremely clever playfulness – private exercises in rhyme and eroticism almost – but often the result is a clarity of tone, a density of meaning and a highly charged poetic fluency, even (or especially) in poems where the rhyme and meter are not so perfectly executed. In his excellent 'Scenes from…' poems ('Scenes from Stendhal,' 'Scenes from Flaubert' etc.), Woods compiles numbered lists of ostensibly unconnected moments, which, when read together, become suggestive, funny and enthralling. In other pieces such as 'Narrative Poem' or 'Drive By,' what appear to be casual observations lead us towards disarmingly evocative lines: "When the bell / for bedtime rings in the asylum, lust // resolves itself and greed is spent like bile." *An Ordinary Dog* is a wide ranging collection, moving from the strong, if expected, rhymes and imagery of the opening poem, to the less expected, exciting repetition of the book's closing stanza:

> The hand of Aeneas
>   veined like a leaf
> the hand of Caravaggio
>   veined like a leaf
> the hand of Darwin
>   veined like a leaf –

Adam Thorpe's *Voluntary* is similarly, if more gently, exemplary in its use of poetic form and music. Particularly striking is his instinct for choosing the right form for the right subject-matter: if a line breaks in order to accommodate a rhyme, it often works to accentuate meaning too. In the poignant 'Home Videos,' for example, the ordinary details of a child's party, once jerkily and haphazardly filmed, become precious but painful indicators of the passing of time: "the slow / lighting of the candles that refuse to light". These are very personal poems, and Thorpe manages to inhabit the poems while remaining outside them, expertly producing vivid pictures of a real life. Packed full of strong feelings, these poems can produce strong feelings in the reader, too. The "Father" poems, such as 'Summing Up,' or 'Clearing Your Study' are especially moving:

> I reach into the bowels of your oldest files:
> what are lives but the illusion that all this
>
> matters? So easily scattered, it slides
> into the third bin-bag, already obese,
>
> in a landslip of receipts (mostly shillings)
> from long-dead stores, more recently-accumulated miles

never used, the special offers milling
with ancient reminders marked, in biro,

Replied to.

In these perfectly pitched pieces even Thorpe's descriptions of the bin-bags are good. *Voluntary* may contain some poems which seem too private and opaque, but this is nevertheless an emotional, vivid and spell-binding collection.

Paul Durcan's *Praise in Which I Live and Move and Have my Being* is a substantial new collection and it too is very personal, regularly addressing friends, relatives or acquaintances in the poems' titles and epigraphs. In contrast to Woods or Thorpe, Durcan's writing seems at first to be quite prosaic, having as its main aim the desire to tell a good story. Yet it is probably in his unsurpassed ability to tell a good story and to tell it so masterfully where we find the deepest of poetic music. Like Kavanagh, Durcan turns local place-names – Sandymount Green, Merrion Row, Grafton Street – into exotic treasures, and in poems such as 'Post-haste to John Moriarty, Easter Sunday, 2007' this becomes a source of melody:

> Joe John Mac an Iomaire of Cill Chiaráin,
> Intoning the lament,
> In the sean nós,
> *Caoineadh na dTrí Mhuire,*
> His hands on his hips,
> The stones of Connemara
> Shining in his cheekbones [...]

As elsewhere in Durcan's work, what is so impressive about this collection is the vision; there are metaphors and refrains here which manage to be both simple and poetically powerful, such as the description of "Old people on their hind legs" in the John Moriarty poem, or the repeated couplet in 'Oaxaca': "*Is not that the chair you sit in? / No, that is the chair I look at.*" If you are lucky enough to have seen Durcan perform his poetry, then you will know what to expect from the best poems here: poems which make you laugh, cry and feel that strange thrill that comes from being in the presence of a truly great and original art.

Tara Bergin is studying at Newcastle University for a PhD on Ted Hughes's translations of János Pilinszky. A selection of her poetry appears in *New Poetries V* (Carcanet, 2011).

On The Move

ALISON BRACKENBURY

J.O. Morgan, *Long Cuts*, CB editions, £7.99, ISBN 9780956735928;
Jean McNeil, *Night Orders*, Smith/Doorstop, £9.95, ISBN 9781906613396;
Carole Bromley, *A Guided Tour of the Ice House*, Smith/Doorstop,
£9.95, ISBN 9781906613310;
John Mole, *The Point of Loss*, Enitharmon, £9.99, ISBN 9781907587047

I have rarely raced through a book as fast as through J.O. Morgan's *Long Cuts*. Here is a long poem you cannot put down: the story of Rocky, "Not mechanic. Toolmaker", mythic maintainer of motorcycles or turbines. If, like me, you missed Rocky's debut, you can still jump aboard the bike.

Morgan is a skilled story-maker. Period details, "sponge pudding" – and Red Guards – enrich without halting his flow. He cross-cuts from Scotland to the Antipodes. Italicised speech captures laconic Rocky, drunken friends and brusque employers. Morgan manufactures muscular free verse, occasionally rhymed. Metres sometimes judder, never stall. Here is Rocky's absorption into engineering:

> the waver of chains, of belts reaching up to the rafters,
>   the dark and the dust, the glow from furnaces.
>
> *Four years for a fitter's apprenticeship.*
> *And that's only if you show promise.*
>
> But the boy is not listening,
> has already begun to pass
> through the sudden shifts in heat
> to clear away the concentrated grime
> collected round the workmen's feet.

Long Cuts pays homage to manual skill. Effort hammers through repetition and rhythm: "Sweat on his forearms, sweat in his hair". Nor are class politics glossed over. Taunted by an officer, Rocky sabotages his ship. If readers think his recurring women pall, Rocky might agree. The most intense lyricism is devoted to machines. A BSA Bantam's blue paint is "midnight, abyssal, leviathan".

Morgan leaves Rocky abruptly in New Zealand – "for a while" – I suspect, joyfully, that the blue BSA Bantam has not reached the end of the road.

> This is where I live now
>   speared by cold fire.
>
> Listing to starboard
>
> new ice
> forming.

The starting points for Jean McNeil's compelling and spacious *Night Orders* were expeditions to both Polar regions. Her introduction explains that her collection, named after a ship's log, honours the Polar "documentary tradition", deploying poetry, printed sources and "prose poetry". McNeil, a novelist, has a powerful gift, in prose, for phrases (Antarctic sun is "a gold axe") and for focused endings: "You will never come this way again". Her poetry has a sustained music, echoing "dead men names: Brabant, Livingston, Biscoe". Her introduction hints at "love and loss", surfacing amongst ice as "dark allure", "the lavish mystery of the other". This love affair with the letter 'l' continues in her luscious litany of ice colours: "dusk opal, albino, rose-grey".

McNeil the literary traveller has an ear as fine as her eye, savouring words from different languages and disciplines: "He catalogues the ice for me – *frazil ice,* / *frost flowers, stambuka*". She is deft with fact: "Orion is reversed". Her descriptions draw the landbound into the monotony and beauty of shipboard life, where the Night Orders lie in "a bronze pool of light".

McNeil avoids preaching, but restraint lends more power to warnings: "We will live on a winterless planet". As her introduction confesses, *Night Orders* ends in anti-climax. But McNeil's readers (like me) may be so absorbed by her rich voyages that they too are left in "fierce unrequited love" for the fantastic, fragile ice.

A gentler experience might be expected from Carole Bromley's first full collection, *A Guided Tour of the Ice House*. But visitors are warned: "We will be underground / longer than you think". Intriguing, alarming, Bromley's poems make excellent travelling companions. She is an irreverent Time Lord. Charlotte Bronte – with "a Laptop" – moans about "that curate": a good joke, darker if readers recall that bearing his child will kill her.

Some of Bromley's boldest journeys are imaginary, including a trip to a "glittering ice-house [...] If I'd been Santa Claus". Her poetry celebrates change. She declares no nostalgia for displays of fifties kitchenware: "it's time to move on". Family poems display astringent honesty: her mother's hatred of Christmas, a nurse seen "spooning in" food to Bromley's father on her final visit. Her lively poems about grandchildren proclaim no guiding wisdom, but "desperate measures, dancing / to Venus in Blue Jeans".

Bromley's records of experience occasionally seem raw and rushed. But her best poems have a beautiful clarity of pattern. In 'Unscheduled Halt', the words "midnight" and "new moon" encircle a brief poem, with love at its centre. Her blank verse, confidently accented, moves energetically:

> the men who
> hacked ice from the ornamental lake
> and dragged it here were starved with cold.

The collection closes with mysterious news, snow-lit:

> So soft,
>
> each flake touched
> the window, as if
> it had never been.

Simplicity startles, in delicate sympathy. Bromley's strong, courageous work can carry readers to unexpected, transforming destinations.

John Mole's *The Point of Loss* begins "How to confront the troll". His travel tips for troll country include "dancing feet". This author of many fine collections follows his own advice. Light syllables run to a radiant close:

> It was too long ago
> to be other than Eden
> when he placed on her table
> the sun and the moon.

'Sparrows and Bamboo' has an ending as exquisite as the "Japanese miniature" it re-creates: "the music they print/ on a sheet of air".

Yet Mole's poems combine toughness with tenderness. Free-wheeling couplets urge the reader to move like a cat: "gigantically grown / and pounce on the unknown". In more spacious lines, a ghost speaks, gently implacable:

"You know that's how it has to be". Endings may last a breath too long. But Mole's shortest lines contain memorable confrontations. John Clare "watches the future drive off / in its shining hatch-backs".

Mole's work treads the finest of lines between joy and loss. He mocks his younger self, the earnest ballroom dancer: "oh my God, the pumps". But his present insights are boldly memorable: "Time disposes. Love is all." He notices how a wet beach "still holds our reflection / against the dark". There are lines in *The Point of Loss* which could turn to proverbs, charms against formless dark. These include his version of Rilke's poem about a woman, almost blind, who "prepares for flight". A journey through Mole's dazzling collection reminds us that poems can not only dance, but fly, in space, through time.

Alison Brackenbury's latest collection is *Singing in the Dark* (Carcanet, 2008). Her eighth collection of poems is due from Carcanet in 2013.

℘

Other Worlds

PASCALE PETIT

Jane Hirshfield, *Come, Thief*, Bloodaxe, £9.95, ISBN 9781852249243;
Jane Griffiths, *Terrestrial Variations*, Bloodaxe, £8.95, 9781852249274;
Amanda Dalton, *Stray*, Bloodaxe, £8.95, 9781852248925;
Tess Gallagher, *Midnight Lantern: New & Selected Poems*, Bloodaxe,
£12, ISBN 9781852249342

The effect of Jane Hirshfield's poems in *Come, Thief*, despite their aphoristic rigour, is rapture. The whole collection is an invitation to the reader to embark with her on close philosophical enquiries but to think through the senses. In 'Invitation', she wonders what an invitation smells like. The olfactory sense is also vivid in 'Rain Thinking', when a house quietens to allow the poet to consider not the sound of rain, but the thought of it. This observation could remain abstract but is firmly anchored in the animal, when she compares the sound to a dog opening and closing his mouth at the thought of a new scent.

A meditation on sell-by-dates and mutability leads to the joyful seizing of the present in 'Perishable, It Said':

> How suddenly then
> the strange happiness took me,
> like a man with strong hands and strong mouth,
> inside that hour with its perishing perfumes and clashings.

Objects of the physical world are stripped down to their essences and listened to as if they have spirit depths. Sometimes her reach is vast, as in 'These Also Once under Moonlight', where she imagines our species looking back at itself as monstrous, transitional but yearning: "Fossils greeting fossils, / fearful, hopeful. / Walking, sleeping, waking, wanting to live".

Emotions can be mediated through objects and therefore made bearable, grief visualised as brown leaves, pain as seawater. As if in defiance of Alzheimer's, a resident of a nursing home responds to the question of how he is with "Contrary to Keatsian joy", despite the voice that urges him to fall from "inside the pear-stem".

Like Hirshfield, Jane Griffiths asks the reader to pay close attention to the visible world and make sense of it. The opening poem, 'Christina's World', after the painting by Andrew Wyeth, sets out this premise. She admits she is attempting the impossible through the persona of the girl crawling among minutely painted grass blades. The subject is Wyeth's neighbour who suffered from polio, but the poem responds to the artist's dilemma: whether to depict human suffering or its transcendence to reach the revelatory:

> If I paint what happened, who'd look at the grass
> again, or trace the shape of what can't be spoken
>
> in the grey-skied space between the house and the barn?

Griffiths's poems in *Terrestrial Variations* are woven with shifting minutiae where nothing is certain, but her gaze is scrupulous. She paints with words as if her colours are still wet, each brushstroke shining with possibility. She has a gift for stunning last lines. In 'Inhabitation', a poem that starts as a play on the word "habit" she makes leaps of association to land on the line, "The house, spun from the whole cloth," holding the poem together with that weave.

Houses and gardens have lives of their own, beyond their inhabitants. In 'After the flood', they have metamorphosed and the residents' reflections speak. Several poems explore the dislocation of moving house. The title poem, 'Terrestrial Variations', asks whether there are multiple variations of our lives, all co-existent. 'Hotel at a Distance' has the enigmatic quality of a Joseph Cornell box, where an architect and his hotel are juxtaposed to create a haunting collage, so that: "The sky's a wrap; he has walked it. / It's like ascending the inner ear."

In contrast to Griffiths, Amanda Dalton starts out with a recognisable world where all seems homely, down-to-earth, smelling "of soap and Co-op bleach", but it's the people in *Stray* who are awry. Whether it's the boy who turns himself into a dog or the orphan twins Gull and Hake, transformed by an accident at sea, or the man who obsessively collects tickets from bus rides, Dalton has the unerring knack of portraying the broken worlds of her characters. She is a hypnotic storyteller who convinces with precise concrete details. The suddenly dead "go on feeling the world / like a man in decent waterproofs".

I heard her read 'Feral' at Lumb Bank and her powerful portrait of a feral dog-child has stayed with me ever since, but the jewel of her long-awaited second collection is the final sequence, 'Lost in Space', which narrates (with hallucinatory accuracy) the process of a mind fracturing under the pressure of a family secret. Dalton's masterful description of Allan remembering the trauma is utterly compelling, as is her evocation of his valiant retreat from it in an escape to the moon as Buzz Aldrin:

> He holds the moon in his head,
> recites it like a prayer to keep the other worlds at bay:
> *The Known Sea, Halo Crater, Sea of Crises, Marsh of Decay.*

Tess Gallagher's *Midnight Lantern: New & Selected Poems* is a bonanza of a book. It spans forty years and includes her moving elegies for her late husband Raymond Carver, her experience of breast cancer, and new poems set in Ireland. But these full-bodied poems are not dark. Even when Gallagher is spare, as in 'Sudden Journey', her generosity of spirit comes through. In interview, she has said: "the most wonderful poems can lift you, almost physically, leaving you to hover above the earth," and this is what her best poems achieve:

> I can drink anywhere. The rain. My
> skin shattering. Up suddenly, needing
> to gulp, turning with my tongue, my arms out
> running, running in the hard, cold plenitude
> of all those who reach earth by falling.   ('Sudden Journey')

In 'Bull's Eye', when her subject is the moment a doctor diagnoses a terminal disease, she grasps it to celebrate life: "an arrow flew out of / the cosmos – *thung!* / Heart's center. Belonging / to everything. That / quick." She is equally adept at tackling subjects such as the Holocaust in 'The Women of Auschwitz', and does so through her experience of chemotherapy and baldness. This big-hearted embrace of important themes does not prevent her from being inventive: 'My Unopened Life', for example, depicts surroundings from a reflection in a spoon, and 'Iris Garden in May' magnifies black irises to Georgia O'Keeffe scale, their corollas the size of cathedrals.

When describing the night before a lover's death, the stars are embedded in his chest like diamonds, as she helps him to "hurl / his heart-beat like the red living fist it was / one more time". The title poem 'Midnight Lantern', from the new, more pastoral poems written in Ireland, reads like a manifesto of her poetics. These poems illuminate everything that is dark and makes it shine with affirmation:

We brave into ourselves each time
we put on our lantern-light
and step out [...]   ('Midnight Lantern')

Pascale Petit's latest collection, *What the Water Gave Me: Poems after Frida Kahlo* (Seren, 2010), was shortlisted for both the T.S. Eliot Prize and Wales Book of the Year, and was a Book of the Year in the *Observer*. She leads poetry courses at Tate Modern.

ℬ

What Age Am I?

DAVID MORLEY

Best New Poets 2011, ed. D.A. Powell, Samovar Press, $11.95,
ISBN 9780976629665;
The Best British Poetry 2011, ed. Roddy Lumsden, Salt, £9.99,
ISBN 9781907773044;
The Salt Book of Younger Poets, ed. Roddy Lumsden and Eloise Stonborough,
Salt, £10.99, ISBN 9781907773105

At the risk of killing this review with its first sentence, let us examine anthology selection methodology. The American *Best New Poets 2011* uses a blind process. Well, not quite blind – a kind of sightlessness with conditions, connections and cash: "Each magazine and [writing] program could nominate two writers... For a small reading fee, writers who had not received nominations could upload two poems... Five readers and the series editors blindly ranked these submissions, sending one hundred and eighty five manuscripts to this year's guest editor, D.A. Powell, who selected the final fifty poems". So: that includes writing programmes, literary editors and poets who can pay up online. D.A. Powell sums up the traits of his generation of poets as "a healthy distrust of pervasive wisdoms and sentiments". Quite. As for *The Best British Poetry 2011*: "These were the poems I felt were best, of all the poems I read". Of those fuddy-duddies troubled by the notion of 'the best': "If it really bothers anyone, a cup of tea and a nap might help".

No effective map of US and UK poetry is offered by either anthology, although one claims to represent "the finest and most engaging poems found in British-based literary magazines and webzines" (so long as you are not publishing in the *London Review of Books* or *Times Literary Supplement*); while the other "cannot suppose we know what poetry will look like in twenty years; perhaps not even what it will look like tomorrow. But this is a glimpse at today: the energy, openness, and chutzpah of the new". This got me thinking about the lifeless linking between 'the new' as equalling 'energy' and 'openness'. Either way, the argument presents a cliché from another age – the birth of Modernism passed through the fantastic prism of Blake's "Energy is eternal delight".

Both books contain very strong and immaculately crafted poems. It is fascinating to observe how much more advanced and various are the prosodic devices of the American poets, and how at ease they are with the longer line and with subjects such as science (by which I mean poems that fully inhabit the subject in language, form and vision). The range and play and clarity of formal devices (even at their boldest and most experimental) are the most enlivening qualities from the US *Best*, while the UK *Best* is a masterclass in imagery and interiority of feeling, vision and voice. It is almost like two languages of feeling and perception are being spoken here, both recognisably English and ultra-poetic, except that one is bouncing off the page into the real world, while the other sings – too much to itself at times? – from a brilliantly crafted icehouse of image. And yet each set of poets could so benefit from reading the other, and learning and blending 'the best' (that is, the most ambitious) qualities of both. Few poets have achieved the transatlantic transformative double-dance of language realised by writers such as Amy Clampitt, Charles Tomlinson, Marianne Moore, Roy Fisher, Elizabeth Bishop, Christopher Middleton, John Ash and Thom Gunn. Are younger British poets reading these fine writers for ideas and strategies or are they too busy reading one another?

A quarter of a century ago I was a younger poet. Charles Tomlinson was my mentor. He banned me from reading contemporary poets! That is one of the reasons five years later I did not feel qualified to edit *The New Poetry* for Bloodaxe Books (although the critical distance helped and I belonged to no cliques). I went ahead with the project because nobody from an older generation appeared to be interested in the generation just before my own – 1963 backwards. That was my first false assumption. I made other gormless assumptions along the way but I hope I learned from my mistakes. The most important thing I learned from editing *The New Poetry* is that excluding poets because of their age was a severe mistake because it presented a phoney picture of artistic development. The criterion did not recognise that a good number of poets, especially women, develop later in life.

Please do not mistake me. I recognise the requirement of a parameter to reduce the number of poets to a workable book – I myself am guilty as charged. I am not griping about the poets chosen in *The Salt Book of Younger Poets*. As a result of reading this anthology I have made a point of reviewing each and every one of these poets online on Facebook and Twitter; I have also made a point of reviewing emerging poets who have published to great effect later in life. Salt is an outstanding, innovative poetry publisher and this project is the sort of thing it does spectacularly well, especially as it follows through with first collections by some of the poets included. However, the editorial decision to use age in order to choose one poet over another poet defies critical comprehension. It does not serve the truth of poetry or the truth of the younger poets in this anthology. It strips their work of context; it restricts variety, especially prosodic range; it creates the sense of an artistic sound-bite.

We do not need to go on reinventing the age-cliché as if it does not matter, and as if the poems written by poets developing in their forties, fifties, sixties and beyond also do not matter. 'Younger poets' can be of any age – or as Patrick Kavanagh wrote in 'Innocence': "What age am I? // I do not know what age I am". This is the dilemma. At its heart this project is about the emergence of a highly talented generation of new writers. What I am asking for might seem idealistic but, in my view, is entirely pragmatic: an anthology that seeks to represent and celebrate emerging poets of any age.

David Morley's most recent books are *Enchantment* (Carcanet, 2011) and *The Cambridge Companion to Creative Writing* (Cambridge University Press, 2012). He is Professor of Writing at Warwick University.

ß

The Spleen, The Blues

ALFRED CORN

Charles Baudelaire, *The Complete Verse*, trans. Francis Scarfe,
Anvil, £10.95, ISBN 9780856464270;
Charles Baudelaire, *Paris Blues: Le Spleen de Paris. The Poems in Prose:
Petits Poèmes en prose*, with La Fanfarlo, trans. by Francis Scarfe,
Anvil, £10.95, ISBN 9780856464294

The late Francis Scarfe began his professional life as a lecturer in French poetry at Glasgow University and served, from 1958 until 1979, as the Director of the British Institute in Paris, meanwhile publishing four books of his own poetry and two books of criticism. These sound like valid credentials for someone attempting to translate Baudelaire. His versions are not miles off the mark, yet they don't seem as close to the original as they might be. The volume of verse comes in dual-language format, with Scarfe's prose translations at the bottom of the page. If he had tried for metrical, rhymed versions, and if we regarded him as an important poet in his own right, then departures from Baudelaire's exact meaning would be acceptable or even laudable. As it is, they exist in an odd limbo, rather too far from the French to serve as a crib, and yet not impressive as poems in English.

I should offer evidence. The sonnet 'Tristesses de la lune' begins with this line: "Ce soir la lune rêve avec plus de paresse". Scarfe's title is 'Sorrows of the Moon', though "sorrow" is darker than "sadness". For the first line he gives: "Tonight, the Moon is more languidly than usual dreaming". Commissioned to provide a prose trot, other translators might say, "This evening, the moon dreams with more indolence". That's a close replica of the original, and, besides, there are several reasons to avoid "languidly". We find no adverbs in the French line; adverbs (in particular, adverbs in –*ly*) weaken a text. Also, Baudelaire farther down uses the noun "langueur", which Scarfe translates "languor", making it appear to be a partial echo of "languidly" in his first line. A word-by-word version wouldn't leave this false impression. Meanwhile, if we judge Scarfe's opening by the standards of poetry, its shaky syntax, its inflated upper-case "Moon", and the absence of any effort to find a counterpart for Baudelaire's alexandrine metre would disqualify it for high praise. I wish Scarfe had been more a poet here, or else not so much. That's not to say the book is useless. It gives the French text of all Baudelaire's poems in an inexpensive volume. If you know French, the English will often save a riffle through the dictionary, but you'll also see things not found in the original.

Readable and informative, Scarfe's introductions to both volumes are also of value. They give us a portrait of Baudelaire as more idealist than realist, more vulnerable than dominating, more harmed than harming. The poet himself believed he was born under an unfavourable star. His sonnet 'Le Guignon' (translated by Scarfe as 'Unluck', but more naturally rendered as 'The Jinx') is an oblique meditation on the topic, using the myth of Sisyphus to suggest the specific gravity of his unhappy destiny. The conclusion echoes Gray's 'Elegy Written in a Country Churchyard', at least, the stanza beginning, "Full many a gem of purest ray serene...". Gray's stanza also prefigures Emerson's 'The Rhodora'. Baudelaire had some English and seems to have read Emerson's first 'Nature' essay, but whether he knew of 'The Rhodora' and Gray's elegy, who can say. What the sonnet implies is that Baudelaire regarded himself as a person comparable to Gray's undiscovered heroes, a treasure condemned to obscurity by the "jinx".

Related to Baudelaire's *guignon* is his fondness for the term "splccn", borrowed from English and in his era meaning not so much "anger" as "melancholy". For the volume of poems in prose, Scarfe has translated *Le Spleen de Paris* as *Paris Blues*. He justifies the choice by noting that "spleen" is no longer used in Baudelaire's sense and that "blues" as a term for "sadness" dates to the eighteenth century. Still, when they see this title, readers who don't often turn to the *OED* are more likely to think of Bessie Smith or 'St Louis Blues' than the Second Empire melancholy Baudelaire laboured to evoke. Note, too, that in the volume of verse, one of *Les Fleurs du mal*'s subsection titles, "Spleen et Idéal", was rendered as "Spleen and Ideal". If "spleen" is an inapt translation of *spleen*, Scarfe has nevertheless used it.

Title aside, the prose poems read well in English, no doubt because prose is easier to comprehend, and therefore to translate. Scarfe says that he allowed himself more freedom in translating the prose, though I don't see that it deviates from the original more often than does his translation of the poems. Perhaps he *felt* freer when doing the prose; it seems relaxed, unfinical. We might assume that Baudelaire's prose poems were in their way as powerful as the poems. Yet they seem bland by comparison, the best proof, a consideration of poems that he demoted to prose for the new purpose. Consider, for example, 'La Chevelure' (Scarfe gives 'Hair' but 'Tresses' is more like it), which in prose becomes 'Un Hémisphere dans une chevelure' ('A Hemisphere in Her Hair'). The prose remaking occasions a sharp falling off of intensity and striking sound. The same can be said for the poems

'L'Invitation au voyage' and 'Le Crépuscule du soir' once they try to manage without distillation and prosody. Not that my estimate, in the year 2012, makes any difference. They are still read and have engendered many imitations, some brilliant. "Translation" can also be applied, figuratively, to literary influence, and Baudelaire is one of the gold-medal influencers. As a bonus for the reader, Scarfe appends the novella *La Fanfarlo*, the only work of fiction Baudelaire ever produced. Not described as a prose poem, the narrative even so has some passages of fine (at times, cynical) writing. I don't recall ever having seen an English version of it, nor do its occasional infelicities ruin the overall impression. These two volumes contain all the Baudelaire the general reader needs, and a little more.

Alfred Corn has published nine books of poetry. This year he is a Visiting Fellow at Clare Hall, Cambridge, and preparing a new version of Rilke's *Duino Elegies*. *Transatlantic Bridge: A Concise Guide to the Differences between British and American English* was published this spring.

℘

Smoke And Mirrors

HANNAH LOWE

Anthony Joseph, *Rubber Orchestras*, Salt, £12.99, ISBN 978-1-84471-81901;
Isobel Dixon, *The Tempest Prognosticator*, Salt, £9.99 ISBN 9781844718252;
Tamar Yoseloff, *The City with Horns*, Salt, £9.99, ISBN 9781844718184;
Katy Evans-Bush, *Egg Printing Explained*, Salt, £9.99 ISBN 9781844718221

These four collections range inventively across time and place, evoking media, myth and history. The Trinidadian Anthony Joseph is by far the most formally experimental, his poems composed, the inlay tells us, using the as yet undefined technique of "liminalism", related (one assumes) to the state of liminality in which fixed identities dissolve in favour of openness and indeterminacy. 'A Ditch of Knives' begins "To be national, liberty is impossible", announcing Joseph's counter-position, to resist confinement through the criss-crossing of spatial and temporal borders.

Rubber Orchestras is the poet and musician's fourth collection, its name taken from a phrase coined by the Surrealist artist Ted Joans. Indeed, the surrealist mode of these poems will be a challenge to some readers, but I enjoyed their psychedelic aesthetic and the bravado of Joseph's writing in which multiple cultures are collided, the narrator journeying across and within poems from African civilisations to the Harlem of the Jazz Age to the past and present day Caribbean. The phrase "I came from" and its variations serve as testimony and refrain, embedding the narrator as witness and agent.

At the heart of the book is calypso, a symbol of the cultural *mestizaje* of Europe and Africa in the Caribbean, thus 'Dimanche Gras' travels between San Fernando, Guyana and Brooklyn, witnessing "Europe kissing calypsonians on the neck" and stressing Europe's defining tendencies, the narrator, a Calypsonian himself, tells us "I go down there again, got married twice, / got written down" while in 'The Vibrant Oases' the musicians "went up river / to get our names back". Poems such as 'Blue Hues' structurally embody musical idioms and linguistically evoke the often licentious worlds of jazz musicians. In 'Riff for Morton' the narrator celebrates the infamy of the pioneering jazz pianist Jelly Roll Morton:

> Jelly you rascal you minstrel you lover you
> bone meat of the creole Caribbean
> – vicious semen, Jelly,
> bake 'em brown and break a banjo across their backs.
> Jelly you blues talker you, the voodoo of your laughter,
> stepping lean in stove pipe suits, down
>   to the very end.

Our interactions with the natural world are a repeated theme in Isobel Dixon's colourful *The Tempest Prognosticator*, in which human sensibilities are frequently interwoven with animal life. Poems travel from the Karoo to the Congo to England, responding to science, history, art and film. The title derives from Dr George Merryweather's 1850 invention for predicting storms based on the activity of twelve leeches kept in separate bottles of rainwater. Here and elsewhere, our use and violation of animals speaks of a lack in us, thus the "the hedge-fund billionaire / the scientists, the artist" are condemned in 'Requiem' concerning the goings on around the purchase of Damien Hirst's shark. Likewise, in 'The Merry Jesters' Rousseau's apes stare out into a gallery with "tragic eyes" at the "zoo-keepers, hunters", who, like us, are "the future selves they fear to see". Animals take multifarious forms, as harbingers, familiars or lovers, as in the humorous, poignant 'You, Me and The Orang-Utan' where the narrator explains her preference for primate to a jilted lover:

"His heart / so filled with care/for every species. And his own, so threatened, rare – / how could I not respond...".

Dixon's poems offer various female subjectivities through what fellow South African J.M. Coetzee's blurb calls her "ventriloquizings of the female self", in poems such as 'Usury' and 'Housewifery'. The language of all the poems is taut, employing surprising images, as in 'Paradox', a treaty on love: "There's no telling what / will make the heart leap, frog- / like, landing with a sloppy plop". Our endeavour in love "makes a mockery / of us, and yet we lie awake / at night and croak and croak for it".

The central section of Tamar Yoseloff's *The City With Horns* is the most engaging, imagining the life of the abstract expressionist painter Jackson Pollock through his own perspective and those of others in his set including O'Hara and Ginsberg, as well as Pollock's mistress and his wife, Lee Krasner. The double-voiced 'Lee Visits the Studio' conjures Pollock's renowned arrogance, Yoseloff's gritty language summoning this fast, lewd world to life:

> She said that we screwed once –
> must have been drunk, she was so ugly
> she was beautiful

The last couplet is in Krasner's voice: "*I said you're sex on legs, / yes*, he said, *I am.*"

The first section of the book is similarly concerned with our disparate perceptions, this time of the city. "You have monuments of your own" the narrator tells the addressee in 'Wish You Were' implying that cities exist only as our subjective versions of them. The passing of time similarly refutes fixed definitions. The city is "No more than smoke and mirrors –" in 'London Particular' in which the narrator's experience of London is severed from that of her father's decades before, this melancholic fragmentation created through sparse and precise phrasing and enjambment.

In an elliptical mode, the poems in Katy Evans-Bush's *Egg Printing Explained* take their cues from high and low culture of the past and present, her language ranging from archaic forms (brilliantly sustained in 'The Love Ditty of an 'eartsick Pirate') to the informal vernacular of pub conversations. This mix of registers is signalled in the opening lines "So I said to Mark, this is no time for more / whimsical gravitas!"

Evans-Bush's formal variation is evident in sonnets, including a terminal "concept sonnet" utilising Pink Floyd's 'Wish You Were Here', and ekphrastic poems such as 'The Fabiola' after Francis Alys's installation of three hundred amateur portraits of the saint of that name. "Happy Fabiolas gaze doll-like all around", are "beautiful, or pretty or plain" and "face the future, / or was it the past?", their variousness echoing the multiple strands and lenses of this book. Elsewhere, characters from myth and history walk in: Nerval with his pet lobster, Hansel, pirates, mermaids. The cultural and linguistic juxtaposition is perhaps paradoxically what makes the book coherent, although some may find its sprawl frustrating.

Like all these poets, Evans-Bush roams across time and space. History repeatedly forces itself on the present, as in 'You're In Bedlam' where the pleading of past inmates are "recorded by the bricks we live among" and in the nostalgic 'Connecticut Postcard' where the use of tense and line-breaks makes time wind and unravel like a yo-yo. Evans-Bush (like Yoseloff) is from the US, living in England. An American childhood of "cicada-scream / summers so bright the leaves were only glare" is evoked here and elsewhere in wistful, cinematic detail.

Hannah Lowe's pamphlet *The Hitcher* is published by *The Rialto*. Her first collection *Chick* will be published by Bloodaxe in 2013.

ℬ

Cargo Of Worlds

STEPHANIE NORGATE

Caroline Carver, *Tikki Tikki Man*, Ward Wood, £8.99, ISBN 9780956896940;
Stuart Henson, *The Odin Stone*, Shoestring, £9, ISBN 9781907356391;
John Gohorry, *A Manager's Dog*, Shoestring, £5, ISBN 9781907356353;
John Weston, *Echo Soundings*, Shoestring, £9, ISBN 9781907356452;
M.R. Peacocke, *Caliban Dancing*, Shoestring, £9, ISBN 9781907356315

Caroline Carver's fourth collection, *Tikki Tikki Man*, opens the door to an "illusory city" built by the "concentrated thoughts / of every mind". Varying landscapes, Jamaican, French, Scottish, Canadian and legendary, become the merging sites of innocence and exploitation after a grieving father entrusts his children to an abusive friend. Carver's imagistic poems with their fragmented yet rhythmic lines form a compelling narrative, bold in its multi-cultural transitions which, for a lesser writer, might lead to dilution instead of intense emotional surprises. Disturbing images of nature, "sea urchins / with black bristles", foreshadow the abuse, juxtaposed by rhythmic objections from the *tikki tikki man*, whose counterpointing voice expresses his own and society's denial. The fabulous, "at night / when everyone had gone to bed we'd pretend we were in a Sultan's island", mixes with the real, "his hands slip slopping with kindness / till they stray over the places / where our breasts will come". Wherever the girls then go, the *tikki tikki man* moves "through walls and doorways / easy as water". Through contact with nature and the ordinary – "as I slip naked into the water I tell Maia / —it's like putting on a dress / made of moonlight" – the women gradually begin to heal. The final poem, 'Postscript', shows that while even pigeon mess can become transformative, "brightening the roses of Isfahan", yet for women, there is ambiguity in burdensome symbolism ("*il faut souffrir pour être belle*"). The journeying sequence takes us through a "cargo of small worlds" which, like Isfahan's "cucumber and melon fields" nurture fruits of friendship and of the imagination as redemptive forces against exploitation.

The Odin Stone by Stuart Henson carries a cargo of worlds too, and is impressive in its range of allusions and use of forms. In the sequence, 'Rilke in Florence', Rilke's creative consciousness becomes a lens through which temporality, painting, light and shade, are observed and celebrated. Financial betrayal of the common good is skillfully etched into the landscape of 'Judas Trees', where "the river is all / pieces of silver" and "the Judas trees / burst into shame". From 'Verlaine in Camden' to variations on Pushkin to the building of a bridge or a metaphysical description of eggshells fallen from the nest, Henson develops dramas that value the personal narrative, the meditative eye (as imaged in the odin stone on the cover), and explore pressures on individual perception whether linguistic, emotional or political. Henson is a witty poet too; a delightful poem, 'The Umbrella', rhythmically echoes the tripping rush of a woman going to meet her lover. Half-rhyme is used in these couplets with brio, avoiding the sentimentality that full rhyme might accord by maintaining the lovers' slight separateness. Henson's work makes literary expectations of its readers, but is never simply showy; form echoes meaning, and the contemporary world and its concerns are vividly present in this enjoyable collection.

John Gohorry's pamphlet, *A Manager's Dog*, is also allusive, taking off from an epigram by Pope. In this traditional satire, with its references to animals, power structures and empty materialism, Fido recounts his master's changing fortunes in a variety of metres. Despite the cutting satire, an appreciation of our dependency on dogs gleams through, "Dogs sift through the debris, their sense / alert to the least sign of life / eyes, nose, and ears attuned perfectly/to the pitch of human survival." The final female villainy might not be missed, though Juvenal would approve; otherwise, as they say of novels, I couldn't put this down.

John Weston's life as a diplomat creates an expectation of engagement with the wider world. Concerns for political freedom are evident in *Echo Soundings* and in 'For Shi Tao, in a Chinese Prison' where poetic syntax effectively emphasises an unfinished sentence. Poets memorialise the soon to be lost, and bee poems have become an urgent vehicle (see Shapcott's and Duffy's recent work) for questioning society's relationship with nature. In the extended sequence, 'Bee Lines', the poem 'Diary' opens with the arresting statement, "Today I became / husband to ten thousand bees", a marriage playfully referenced by the bee-keeper's veil. Through Weston's tender record of his year's husbandry, we learn much about the actual keeping of bees, as well as about associated language and legends. Throughout, the word "colony" reminds us that the bees' cooperative work is one model for living, yet bees require charms and magic, tricks of smoke, not simple governance. Weston wonders why "in the old photos / beekeepers went without veils" and in that wondering we see the keeper working in the face of a lost cohesion with the natural world. The honouring of occasions, an anniversary poem, an epithalamium, sonnets, sestinas, villanelles, linguistic games, ekphrastic poetry and even (that rare thing) a good poem about writer's block, all show Weston engaging with the full range of poetic curiosities with knowledge and yet a modest voice. Sometimes, though, the economic spirit of Shi Tao's work in Weston's striking translations could hold more sway.

In *Caliban Dancing*, M.R. Peacocke meditates on the passing of time. She straps on the day "like a first time on skates" and feels herself "wobble off", somehow making transience visible. In 'The Visit' and 'Thirteenth Night', Peacocke subtly reinvents the annunciation, plays on the return of the magi as a lengthy contemporary homecoming. In other poems, she gives voice to Caliban who recognises the damage done to him, "Let them not come again which say / orders is orders so know your place / we learn you behave". Caliban critiques the ambitions of writing, "how he grew little writing himself / on water   Dash and dot / coming to nothing" in contrast to nature's immediacy. These poems and several others employ spaces instead of punctuation, suggesting wariness about closed authoritarian statements. 'The Wintering Lark' is a perfect poem which questions how to live in old age, "A den of grass, a lair of stems; / And what does he eat, the lark in winter / crouched in his breath, his blown husk / of feather, heart like a seed?" The powerfully focused metaphysical 'Jug', lipped in shape, plays on notions of the vessel, body and soul. 'An Inventory of Silence' engages with solitariness, connecting single silences to the world's history. Though frequently the concern is mortality, for instance, 'One Night in Winter' gathers the sounds of a hospital ward, this book explores the endeavour to live in present moments and ends with an image of survival "how sap still forces / bubbles out of a clipped branch". Celebratory, mournful, imaginative, Peacocke's vivid musical poems, more resonant than they seem at first, provoke us to look afresh at the world and our place in it.

Stephanie Norgate's second collection, *The Blue Den*, is due from Bloodaxe in autumn 2012.

☙

Full Bloods And Wunderkinds

PETER FINCH

Graham Mort, *Cusp*, Seren Books, £8.99, ISBN 9781854115485;
Zoë Brigley, *Conquest*, Bloodaxe, £8.95, ISBN 9781852249304;
Merion Jordan, *Regeneration*, Seren, £8.99, ISBN 9781854115553

How long does it take for a nation's culture to mature? In the case of English language writing in Wales, a little longer than many had hoped. What was once called Anglo-Welsh Literature spent most of the twentieth-century bumping from one flowering to the next, throwing up any number of wunderkinds – Edward Thomas, Dylan Thomas, R.S. Thomas, Rhys Davies, Gerard Manley Hopkins, Alun Lewis. The problem was that while half of these poets moved out of Wales as fast as they could, the rest merely had a circumstantial connection with the country and were not, as some native Americans designate themselves, full bloods.

The advent of a modicum of self-determination, with the arrival of the elected Welsh Assembly Government in 1998, has given the country a new feeling of identity, purpose and self-worth. There is, it transpires, a difference between the cultures east and west of the border. Not a huge one, but an identifiable one nonetheless. The new government made early moves to boost its support for culture.

Anglo-Welsh literature's late twentieth century follow on, the-doesn't-really-trip-off-the-tongue Welsh Writing In English, has benefitted considerably. It's studied in the universities, promoted by the media and funded by the Welsh state's cultural machine. If it is not exactly loved by the people at large, then at least it is recognised as something Wales now owns.

Much of the responsibility for seeing poetry in English in Wales through the past thirty years has fallen on the editorial desk of the small Bridgend-based publisher Seren Books. Begun as an offshoot of the journal *Poetry Wales*, Seren has cast its net wide. Robert Minhinnick, Duncan Bush, Gillian Clarke, Sheenagh Pugh, Dannie Abse, Alun Lewis, Fiona Sampson and Pascale Petit are some of its stars.

Being Welsh in Wales is a fine thing but elsewhere it can be a difficult identity to shoulder. As a press Seren has fought its corner well but has found itself all too often marginalised. Selling Wales is such a hard thing. Who wants to know? To counter the tendency, Seren has tried to leaven its roster with the occasional poet from elsewhere. A broad list is a stable list, that's the theory. The problem is that incoming non-Welsh poets often end being thought of as part of the Welsh canon. As bards with a heritage of wrecked industry, sheep and traditional poetry behind them.

One such is Graham Mort. No Welsh blood in his veins, as far as I can discern. No Welsh residency. No Welsh subject matter in his well-turned verse. He's been a Seren author for a while. It has published his poetry, his short fiction, and brought out his new and selected verse. The work in his latest, *Cusp*, is as tightly controlled and well-observed as its predecessors.

Mort is not a people's poet. His poetry has no immediacy, belly laughs, or political diatribe. It is fluid and dense, a rich verse drawn from his observing of the natural world. Mort is out there, deep among the fells, following the becks as they course down their gullies. "The dove's cry comes / again, through the flood's / garbled pronunciations / pouring from the watershed's / ridge to the arched spine / of the river bridge, deepening / with each moment of / rain, each drenched syllable / deliquescing on its tongues."

Cusp is rich, over rich almost, in such stuff. It's the world beyond him that he depicts. Mort rarely lets his personal self intrude. But there are successful poems about IVF treatment and of catheters and hospitals to sweeten the mix.

The book finishes with a twenty-page tour-de-force, 'Electricity', where Mort, master of form and functionality, is at the top of his powers. It's a poem that sparks its way through world knowledge and its consequences, electricity as the grand metaphor, "call me the turning worm at the heart of matter / I'm everywhere..." It held me.

In contrast to Mort, Zoë Brigley is Welsh. She comes from Caerphilly. Her new collection, *Conquest*, however, is published by Bloodaxe and, apart from in one poem rich with place name, betrays little Welsh connection. Engaging with *Conquest* requires the reader to possess some cultural foreknowledge and to do a little work. You'll need, for example, to know something about the Brontës, all three of them, and what they did to get much from 'My Last Rochester', the book's first section. For its third, 'The Lady And The Unicorn', you'd be better off having viewed the tapestries at the Musée de Cluny, Paris, on which the poems are based.

Brigley is adept at the manipulation of form to often spectacular effect. She makes the found poem appear to be the result of long night wrestle with structure in 'Arches', her poem of American place. She causes the double sestina to read like something out of the OULIPO movement in 'Lady, Lion, Unicorn'. At heart, though, this is book about "women questing to rediscover their own desire". Do they find it? I'm not sure.

Retreading the *Mabinogion*, that compendium of ancient Welsh tales made famous in Charlotte Guests's nineteenth-century translation, has become fashionable of late. Merion Jordan's "re-imagining" is presented in arch post-modernist style as a two books printed back-to-back. It's as if they were English and Welsh translations. *Regeneration* splits into 'The White Book' and 'The Red Book', echoing the stories' sources. Here, in tangential takes, Jordan reassembles elements of both the Welsh tales of the Mabinogi and the myths of King Arthur. The Dark Ages returned. World of sword and sorcery. Magic and power.

It's a logical approach for, in the minds of many, the two are irrevocably melded, and represent Britain simultaneously as it never was and as it always will be. Jordan's trump card is to scatter his text with personal footnotes that wind him and his bilingual working-class history into the heart of the myth. It's an invigorating and captivating approach. I couldn't put it down.

Peter Finch is a poet, psychogeographer and literary entrepreneur living in Cardiff. His poetry has most recently been collected in *Selected Later Poems* and *Zen Cymru*, both published by Seren.

ℬ

Psychic Force:
Gwyneth Lewis's *Clytemnestra*

Produced by Sherman Cymru, Cardiff, 18 April-5 May 2012

GRAHAME DAVIES

It is the near future. Civilisation, if you can call it that, is confined to strongly-defended compounds threatened by lawless 'ferals' outside the gates. Oil is scarce, food is becoming scarcer, and the isolated city-states wage war and form unholy alliances to secure the sustenance they need. Survival is the primary virtue, and when hunger threatens bonds of kinship and common humanity are strained to breaking point.

King Agamemnon has departed on a military expedition intended to secure his community's future, but has made a pact with barbaric allies which had necessitated the sacrifice of his daughter Iphigenia. His wife, the play's eponymous protagonist, rules in his absence, but is becoming ever more consumed by the need to avenge her daughter. It soon becomes apparent that the enemies outside the walls are not the most destructive threats in this dystopia: within the walls, and within the characters' minds, inchoate, instinctive, primitive forces, the Furies, strain for expression, for fulfilment and for revenge.

That is the scenario of *Clytemnestra*, the first play by one of Wales's most distinguished and versatile poets, Gwyneth Lewis, which was chosen as the first commissioned English-language production at the Sherman Cymru theatre in Cardiff, following its re-opening after a six-and-a-half million pound refurbishment. The project began life when Lewis, asked to consider reworking Aeschylus's *Oresteia*, decided that instead of rewriting one of the existing works, she would create a new full-length piece around the character of Clytemnestra. Three years later, the work has been produced by Sherman Cymru's in-house company under the bold direction of Amy Hodge, with Jaye Grifffiths in the title role and Agamemnon played by Nick Moss.

While Lewis has selected from among the many extant versions of the myth of Clytemnestra, and has brought her own approach to the narrative, the real originality of this production lies in its use of language, and in the way Lewis's Furies are conceived. Those who attend Cassandra, the Trojan seer captured by Agamemnon, are described in Lewis's notes as "Proto-Furies... undeveloped pre-linguistic entities". Although existing to avenge Cassandra's family, who have been killed by Agamemnon, they are unable to use her as their vehicle, as she denies them her co-operation. However, the Furies who accompany Clytemnestra and Aegisthus – the ruler of a neighbouring compound who again has lost his family to Agamemnon, and who forms an expedient attachment to the lonely Clytemnestra – are altogether more embodied, able to speak, eerily anticipating and mimicking their humans' words, and gradually gaining the power to drive their actions, as the humans assent ever more fully to the principle of revenge of which the Furies represent the "ancient psychic force".

It is hard not to assume that Lewis's own well-publicised experience of depression has informed these compelling, disturbing creations. Brought to life by Nia Gwynne and Adam Redmore, these entities are acted with uncannily distorted voices and with uncomfortably contorted movements that mirror the deliberately strained diction Lewis has given them. The play would be worth seeing if only for the experience of seeing the interplay of intention and instinct embodied in this striking way.

However, there is a great deal more to this production, too. As might be expected from an author whose alternating volumes in her native Welsh and her early-acquired English have each raised the bar for her successive works, Lewis's vivid poetic ability is much in evidence. Cassandra, played by Kezrena James, calls the insistent voices of the Furies "the blood choir". Asked by Agamemnon whom she is addressing, she replies: "The ones who want us in the light to do / Their dirty work, the vengeance mothers". Here is Aegisthus, played by Jonah Russell, speaking to Cassandra's daughter, Electra, played by Rhian Blythe: "There is a force behind language. / It never forgets. It burrows its way / Into your mind, like a spider's eggs / Laid in the brain, a nest / Of ideas that crawl out at night."

Such intensity is balanced by more naturalistic dialogue. The images in Lewis's poetry are often threaded upon a composed and assured conversational register, and she therefore shows herself equally convincing in the more workaday dialogue of the Chorus who provide an earthy, and occasionally darkly comic, commentary on the main events.

The bonus of having Lewis's entire text within the programme is that it allows the audience to appreciate, at leisure, the craft of this fine poet in this, her first, and surely not last, venture into the world of the stage.

Grahame Davies is a winner of the Wales Book of the Year award. His latest book of poetry, *Lightning Beneath the Sea*, appears from Seren Books this year.

THE GEOFFREY DEARMER
PRIZE 2011

JUDGE: MONIZA ALVI

Each of the twenty entries for the Geoffrey Dearmer Prize 2011 was highly distinctive, as well as a poem of quality. This made the judging a painful process; at times it really did feel like a balloon debate in which one had to decide whether to keep on board the zebra or the hummingbird. The qualities of the winning poem, Denise Saul's 'Leaving Abyssinia', revealed themselves to me slowly, but certainly. Haunting, dreamlike and atmospheric, it's exact, yet hard to pin down. There's a freight of unspoken questions, as to the who, the why, the when, and even the where of the situation – nothing is absolute. The poem could be viewed as a response to Cavafy's 'Ithaka', but a departure, rather than a travelling towards. It is set in the past (Abyssinia), yet it feels personal and present: "I cannot hear what grandfather shouts / from the pier". It is imbued with myth. Robert Graves's *Greek Myths* is a necessary book for the voyage, and Odysseus features, though as an absence. The resounding ending "all the sun returns to the underworld", while conclusive, is suggestive of an ongoing pattern, the battle of the forces of light and dark. The uncertainty is palpable, conveyed as it is by the fog, the windowless cabin and the sense of not knowing whether there's any movement, whether this really is a journey, a journey for the self, as well as for the ship ("unable to tell if I'm moving or not"). The experience feels momentous, the lines measured, but urgent, the sound-patterning quietly emphatic, as in the centrally placed "fishermen / who return boats at sunset hail / those who sail theirs out at dawn".

Abyssinia – and yet the Englishness of *Oxford English Reference Dictionary*, and an indication of the wider world, again in English, with Frances Burnett's *World Dress*. The poem seems to speak, obliquely, of migration, of moving from one life to another, of cultural and colonial history. And it implies spiritual travel, the journey into the unknown one may be compelled to make: "*compelled... to change / that state by forces impressed upon it*". 'Leaving Abyssinia' has beauty in its mystery and clarity, its largeness and its containment.

Other striking poems included Malika Booker's 'Sue Speaks To Me In The Swan Room' with its wonderfully flamboyant imagery. The one-time theatre-going women are now "old parrots, who have lost their flair", but

"Back then we were red / breasted robins; bright Dolly, chirpy Chrissie, / flighty Stella and me". A A Marcoff's 'land, & the river' is passionately observant and transfigurative: "sunrise gathers / its collaborations, / its citrus / potency". Caroline Clark's agile, finely-wrought 'What Is The Word – ' conveys so much of language and languages and of miscarriage itself – its seven lines culminating in: "The English / applies restraint. A thing mislaid, / mistake. A stately horsedrawn flourish. / Miscarriage, o, how ravishingly slow." It is a remarkable distillation, and a strong contender.

The Geoffrey Dearmer Prize is awarded annually to the best *Poetry Review* poem written by a poet who doesn't yet have a full collection. It is funded through the generosity of the Dearmer family in honour of the poet Geoffrey Dearmer, who was a Poetry Society member.

Judge Moniza Alvi's most recent collections are *Europa* (Bloodaxe, 2008) and *Homesick for the Earth,* versions of poems by Jules Supervielle (Bloodaxe, 2011).

ℬ

Denise Saul was born in London. She is a poet and fiction writer. Her *White Narcissi* (Flipped Eye Publishing) was Poetry Book Society Pamphlet Choice for Autumn 2007. Denise's poetry has been published in a variety of US and UK magazines and anthologies. In 2008 she was selected for The Complete Works, a two-year mentoring and development programme for ten advanced black and Asian poets in the UK. The programme was organised by Spread The Word and Denise's mentor was John Stammers. Her *House of Blue* (published by Rack Press) is a PBS Pamphlet Recommendation for Summer 2012.

Denise views poetry as an exploration of the self. "It is through this process that the poet experiences a complete oneness with the universe. It is also a journey of the hero/ine which involves initiation, transformation and formation," she says.

Denise Saul
Leaving Abyssinia

A foghorn sounds: I notice the distance
between houses and the shore
as the ship pulls away from a pillar:
strata of limestone, clay and granite.
A wall of fog drifts towards the coast;
gulls peck at moss behind a stone ledge.
I sit in a cabin without windows,
unable to tell if I'm moving or not.

I cannot hear what grandfather shouts
from the pier – goodbye – perhaps.
The wind billows in the smell of mackerel.
At night, nothing is certain when I leave
this land where morning and night
come so close together that fishermen
who return boats at sunset hail
those who sail theirs out at dawn.

An hour later, the clock ticks the same way.
It's 1.30 a.m. and as I'm still
awake, I light an oil-lamp to read a book.
I packed books which were needed:
World Dress by Frances Burnett,
Oxford English Reference Dictionary
and *Greek Myths* by Robert Graves.

Here, there is *no* Odysseus to let passengers know
that the ship's motion is 'uniform'.
I recall that phrase from a lesson at school
as that formula was rote learnt:
every body continues... in its state
of rest or of uniform motion... in a right line
unless it is compelled... to change
that state by forces impressed upon it.
I leaf through the story of the Clashing Rocks;
all the sun returns to the underworld.

CONTRIBUTORS

James Aitchison has published six collections of poems, most recently *Foraging: New and Selected Poems* (Worple Press). **Gillian Allnutt** is currently putting together her eighth collection of poetry. **Simon Armitage** is curator of Poetry Parnassus. *Walking Home*, an account of his poetic journey along the Pennine Way, is published in July 2012. **Elizabeth Barrett**'s most recent collection is *A Dart of Green and Blue* (Arc Publications, 2010). **Paul Batchelor**'s first collection, *The Sinking Road*, was published by Bloodaxe in 2008. **Fiona Benson**'s work appeared in *Faber New Poets 1*. She is currently working on her first collection. **Peter Bland** is currently living in New Zealand where he has just won the Prime Minister's Award for his poetry. **Rosie Breese** is a poet and teacher currently living in Cambridge, UK. **Carole Bromley**'s first collection, *A Guided Tour of the Ice House*, was published by Smith/Doorstop in 2011. **Hayley Buckland** is a PhD student working towards a first collection. **Dan Burt** lives and writes in London and at St. John's College, Cambridge, of which he is an Honorary Fellow. **Anthony Caleshu** is the author of two books of poems, most recently *Of Whales* (Salt, 2010). **Tim Cockburn**'s pamphlet *Appearances in the Bentinck Hotel* is published by Salt. **Terese Coe**'s poems and translations have appeared in many magazines including *The Times Literary Supplement* and *New American Writing*. **Julia Copus**'s *The World's Two Smallest Humans* is published by Faber in July, and is a Poetry Book Society Recommendation. **Emily Critchley**'s *Selected Writing: Love / All That / & OK* was published by Penned in the Margins in 2011. **Joe Dresner** was born in Sunderland in 1987 and now lives and works in London. **Frank Dullaghan**'s second collection, *Enough Light to See the Dark*, was published earlier this year by Cinnamon Press. **Adam Elgar** is translating the fiction of Alessandra Lavagnino and the sonnets of Gaspara Stampa. **Carrie Etter**'s latest collection is *Divining for Starters* (Shearsman, 2011). She edited *Infinite Difference: Other Poetries by UK Women Poets* (Shearsman, 2010). **Angela France**'s latest book is *Lessons in Mallemaroking* (Nine Arches Press, 2011). **Allison Funk** is the author of four books of poems, most recently *Forms of Conversion*. **Gill Gregory**'s collection, *In Slow Woods*, is published by Rufus Books (London/Toronto, 2011). **Philip Gross** won the T.S. Eliot Prize for *The Water Table* (2009). **Robert Hamberger**'s most recent collection is *Torso* (Redbeck, 2007). **W.N. Herbert** co-edited *Jade Ladder: an anthology of contemporary Chinese poetry* (Bloodaxe, 2012) with Yang Lian. His next collection, *Omnesia*, will appear in 2013. **Michael Hulse**'s most recent collection is *The Secret History* (Arc, 2009). **Helen Ivory**'s fourth collection, *Waiting for Bluebeard*, is due from Bloodaxe in 2013. **A.B. Jackson** is the author of *Fire Stations* (Anvil, 2003, winner of the Forward Prize for Best First Collection) and *Apocrypha* (Donut Press, 2011). **Carolyn Jess-Cooke**'s second poetry collection, *Motherhood as an Orange*, is due out next year. **Martha Kapos**'s most recent collection, *Supreme Being* (Enitharmon, 2008), was a PBS Recommendation. **Philip Knox** lives in Oxford, where he studies medieval literature. **Anja König**'s work has appeared in the UK and the US. **Richard Lambert**'s first collection, *Night Journey*, is forthcoming from Eyewear Publishing. **Lorraine Mariner**'s collection, *Furniture* (Picador, 2009), was shortlisted for the Forward Prize for Best First Collection and the Seamus Heaney Centre Poetry Prize. **Christopher Middleton** lives in the US but is arguably the last remaining British modernist. **Kate Miller** has work in *Best British Poetry 2011* (Salt). **Geraldine Monk**'s latest collection, *Lobe Scarps & Finials*, was published by Leafe Press in 2011. **Theresa Muñoz**'s pamphlet *Close* was published this year by HappenStance Press. **Daljit Nagra**'s latest collection is *Tippoo Sultan's Incredible White-Man Eating Tiger-Toy Machine!!!* **Vivek Narayanan**'s second book of poems, *Life and Times of Mr. S*, is due from HarperCollins India in November. **Jeremy Noel-Tod** is currently revising *The Oxford*

Companion to Modern Poets. **Andrew Pidoux**'s *Year of the Lion* is available from Salt Publishing. **Adam Piette** is the author of *Remembering and the Sound of Words* (OUP, 1996) and co-editor of *Blackbox Manifold*. **Kate Potts**'s first full-length collection is *Pure Hustle* (Bloodaxe, 2011). **Shazea Quraishi**'s pamphlet, *The Courtesan's Reply*, is due from Flipped Eye Publishing in July 2012. **Maurice Riordan**'s next collection, *The Water Stealer*, will be published by Faber in 2013. **Peter Robinson**'s latest collection, *The Returning Sky* (Shearsman Books), was a PBS Recommendation for Spring 2012. **Simon Royall**'s poems have appeared in *The Rialto* and *Magma*. **Stewart Sanderson** is a research student in Scottish literature at Glasgow. **Owen Sheers**'s *Pink Mist*, a 15,000-word verse drama about three friends who go to fight in Afghanistan, was broadcast on BBC Radio 4 across five nights in March. **Tara Skurtu** has received an Academy of American Poets Prize and publishes widely in the USA. **Jon Stone**'s collection *School of Forgery* (Salt, 2012) is a PBS Summer Recommendation. **Matthew Sweeney**'s new collection, *Horse Music*, is forthcoming from Bloodaxe in early 2013. **Angela Topping**'s *Paper Patterns* is forthcoming from Lapwing this year. **Siriol Troup**'s *Beneath the Rime* was published by Shearsman in 2009. **Wendy Videlock**'s collection, *Nevertheless*, is available from Able Muse Press. **Christian Ward**'s first collection, *The Moth House*, is published in October by Valley Press. **Tom Warner**'s poetry appeared in *Faber New Poets 8*. He has won a Gregory Award and the Plough Prize. **John Wheway** is currently completing a new collection of poems on psycho-sexual themes. **Susan Wicks**'s most recent collection, *House of Tongues*, was a PBS Recommendation. **Anna Woodford**'s debut collection *Birdhouse* (Salt, 2010) was a winner of the Crashaw Prize. **Dan Wyke**'s latest publications are *Waiting for the Sky to Fall* (Waterloo Press) and pamphlet *Spring Journal* (Rack Press). **Jane Yeh**'s second collection, *The Ninjas*, is forthcoming from Carcanet this November.

POETRY PARNASSUS

The UK's largest gathering of world poets

26 JUNE – 1 JULY

THE WORLD POETRY SUMMIT CONFERENCE
FINDING POETRY'S PLACE IN THE WORLD

TUESDAY 26 JUNE

A global gathering of leading directors, publishers and writers contemplate poetry's relationship with the worlds of innovation, money, digital progress, translation and more.

In Partnership with the British Council

Queen Elizabeth Hall, 9.30am – 6.30pm £35

BOOK NOW: 0844 847 9910

SOUTHBANKCENTRE.CO.UK/
POETRYPARNASSUS

LOTTERY FUNDED | ARTS COUNCIL ENGLAND
Supported using public funding by

SOUTHBANK CENTRE

FESTIVAL OF THE WORLD

MasterCard